SCIENCE PRIMERS,

edited by

PROFESSORS HUXLEY, ROSCOE,

and

BALFOUR STEWART.

VI.

PHYSIOLOGY.

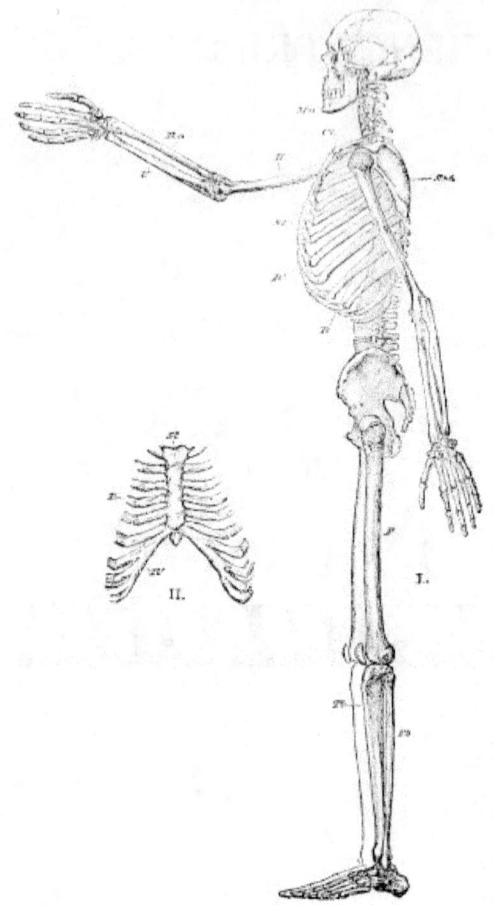

EXPLANATION OF THE PLATE.
FIG. I.—THE HUMAN SKELETON IN PROFILE.

Mn.	The Mandible or Lower Jaw.	
St.	The Sternum.	
R.	The Ribs.	—In the Thorax.
R´.	The Cartilages of the Ribs.	
Scp.	The Scapula, or Shoulder Blade.	
Cl.	The Clavicle, or Collar Bone.	
H.	The Humerus.	
Ra.	The Radius.	—In the Arm.
U.	The Ulna.	
F.	The Femur.	
Tb.	The Tibia.	—In the Leg.
Fb.	The Fibula.	

FIG. II.
A front view of the Sternum, *St.*, with the Cartilages of the Ribs, *R´.*, and part of the Ribs themselves *R.*

Science Primers.

PHYSIOLOGY.

BY

M. FOSTER, M.A., M.D., F.R.S.,

FELLOW OF TRINITY COLLEGE, CAMBRIDGE.

WITH ILLUSTRATIONS.

PREFACE.

THIS Primer is an attempt to explain in the most simple manner possible some of the most important and most general facts of Physiology, and may be looked upon as an introduction to the Elementary Lessons of Professor Huxley.

In my descriptions and explanations I have supposed the reader to be willing to handle and examine such things as a dead rabbit and a sheep's heart; and written accordingly, I have done this purposely, from an increasing conviction that actual observation of structures is as necessary for the sound learning of even elementary physiology, as are actual experiments for chemistry. At the same time I have tried to make my text intelligible to those who think reading verbal descriptions less tiresome than observing things for themselves.

It seemed more desirable in so elementary a work to insist, even with repetition, on some few fundamental truths, than to attempt to skim over the whole wide field of Physiology. I have therefore omitted all that relates to the Senses and to the functions of the Nervous System, merely just referring to them in the concluding article. These the reader must study in the "Elementary Lessons."

M. FOSTER.

SCIENCE PRIMERS.

PHYSIOLOGY.

INTRODUCTION. § I.

1. Did you ever on a winter's day, when the ground was as hard as a stone, the ponds all frozen, and everything cold and still, stop for a moment, as you were running in play along the road or skating over the ice, to wonder at yourself and ask these two questions:—"Why am I so warm when all things around me, the ground, the trees, the water, and the air, are so cold? How is it that I am moving about, running, walking, jumping, when nothing else that I can see is stirring at all, except perhaps a stray bird seeking in vain for food?"

These two questions neither you nor anyone else can answer fully; but we may answer them in part, and the knowledge which helps us to the answer is called **Physiology**.

2. You can move of your own accord. You do not need to wait, like the boughs or the leaves, till the wind blows upon you, or, like the stones, till somebody stirs you. The bird, too, can move of its own accord, so can a dog, so can any animal as long as it is alive. If you leave a stone in any particular spot, you expect to find the stone there when you come to it again a long time afterwards; if you do not, you say somebody or something has moved it. But if you put a sparrow or mouse on the grass plot, you know that directly your back is turned it will be off.

All animals move of themselves. But only so long as they are alive. When you find the body of a snake on the road, the first thing you do is to stir it with a stick. If it moves only as you move it, and as far as you move it, just as a bit of rope might do, you say it is dead. But if, when you touch it, it stirs of itself, wriggles about, and perhaps at last glides away, you know it is alive. Every living animal, of whatever kind, from yourself down to the tiniest creature that swims about in a little pool of water and cannot be seen without a microscope, moves of itself. Left to itself, it moves and rests, rests and moves; stirred by anything, away it goes, running, flying, creeping, crawling, or swimming.

Something of the kind sometimes happens with lifeless things. When a stone is carefully balanced on the top of a high wall, a mere touch will send it toppling down to the ground. But when it has reached the ground it stops there, and if you want to repeat the trick you must carry the stone up to the top of the wall again. You know the toy made like a mouse, which, when you touch it in a particular place, runs away apparently of its own accord, as if it were alive. But it soon stops, and when it has stopped you may touch it again and again without making it go on. Not until you have wound it up will it go on again as it did before. And every time you want it to run you must wind it up afresh. Living animals move again and again, and yet need no winding up, for they are always winding themselves up. Indeed, as we go on in our studies we shall come to look upon our own bodies and those of all animals as pieces of delicate machinery with all manner of springs, which are always running down but always winding themselves up again.

3. You are warm; beautifully warm, even on the coldest winter day, if you have been running hard; very warm if you are well wrapped up with clothing, which, as you say, keeps the cold out, but really keeps the warmth in. The bed you go to at night may be cold, but it is warm when you leave it in the morning. Your body is as good as a fire, warming itself and everything near it.

The bird too is warm, so is the dog and the horse, and every four-footed beast you know. Some animals however, such as reptiles, frogs, fish, snails, insects, and the like do not seem warm when you touch them. Yet really they are always a little warm, and some times they get quite warm. If you were to put a thermometer into a hive of bees when they are busy you would find that they are very warm indeed. All animals are more or less warm as long as they are alive, some of them, such as birds and four-footed beasts, being very warm. But only so long as they are alive; after death they quickly become cold. When you find a bird lying on the grass quite still, not stirring when it is touched, to make quite sure of its being dead you feel it. If it is quite cold, you say it has been dead some time; if it is still warm, you say it is only just dead—perhaps hardly dead, and may yet revive.

4. You are warm, and you move about of yourself. You are able to move because you are warm; you are warm in order that you may move. How does this come about? Just think for a moment of something which is not an animal, but which is warm and moves about, which only moves when it is warm, and which is warm in order that it may move. I mean a locomotive steam-engine. What makes the engine move? The burning coke or coal, whose heat turns the water into steam, and so works the piston, while at the same time the whole engine becomes warm. You know that for the engine to do so much work, to run so many miles, so much coal must be burnt; to keep it working it must be "stoked" with fresh coal, and all the while it is working it is warm: when its stock of coal is burnt out it stops, and, like a dead animal, grows cold.

Well, your body too, just like the steam-engine, moves about and is warm, because a fire is always burning in your body. That fire, like the furnace of the engine, needs fresh fuel from time to time, only your fuel is not coal, but food. In three points your body differs from the steam-engine. In the first place, you do not use your fire to change water into steam, but in quite a different way, as we shall see further on. Secondly, your fire is a burning not of dry coal, but of wet food, a burning which although an oxidation (Chemistry Primer, Art. 5) takes place in the midst of water, and goes on without any light being given out. Thirdly, the food you take is not burnt in a separate part of your body, in a furnace like that of the engine set apart for the purpose. The food becomes part and parcel of your body, and it is your whole body which is burnt, bit by bit.

Thus it is the food burning or being oxidized within your body, or as part of your body, which enables you to move and keeps you warm. If you try to do without food, you grow chilly and cold, feeble, faint, and too weak to move. If you take the right quantity of proper food, you will be able to get the best work out of the engine, your body; and if you work your body aright, you can keep yourself warm on the coldest winter day, without any need of artificial fire.

5. But if this be so, in order to oxidize your food, you **have need of oxygen**. The fire of the engine goes out if it is not fed with air as well as fuel. So will your fire too. If you were shut up in an air-tight room, the oxygen in the room would get less and less, from the moment you entered the room, being used up by you; the oxidation of your body would after a while flag, and you would soon die for want of fresh oxygen (see Chemistry Primer, p. 14).

You have, throughout your whole life, a need of fresh oxygen, you must always be breathing fresh air to carry on in your body the oxidation which gives you strength and warmth.

6. When a candle is burnt (Chemistry Primer, p. 6) it turns into carbonic acid, and water. When wood or coal is burnt, we get ashes as well. If you were to take all your daily food and dry it, it too would burn into ashes, carbonic acid, and water (with one or two other things of which we shall speak afterwards).

Your body is always giving out carbonic acid (Chemistry Primer, Exp. 7). Your body is always giving out water by the lungs, as seen when you breathe on a glass, by the skin, and by the kidneys; and we shall see that we always give out more water than we take in as food or drink. Your body too is daily giving out by the kidneys and bowels, matters which are not exactly ashes, but very like them. We do not oxidize our food quite into ashes, but very nearly; we burn it into substances which are no longer useful for oxidation in the body, and which, being useless, are cast out of the body as waste matters.

The tale then is complete. By the help of the oxygen of the air which you take in as you breathe, you oxidize the food which is in your body. You get rid of the water, the carbonic acid, and other waste matters which are left after the oxidation, and out of the oxidation you get the heat which keeps you warm and the power which enables you to move.

Thus all your life long you are in constant need of oxygen and food. The oxygen you take in at every breath, the food at every meal. How you get rid of the waste matters we shall see further on.

If you were to live, as one philosopher of old did, in a large pair of delicate scales, you would find that the scale in which you were would sink down at every meal, and gradually rise between as you got lighter and hungry. If the food you took were more than you wanted, so that it could not all be oxidized, it would remain in your body as part of your flesh, and you would grow heavier and stouter from day to day; if it were less, you would grow thinner and lighter; if it were just as much as and no more than you needed, you would remain day after day of exactly the same weight, the scale in which you sat rising as much between meals as it sank at the meal time.

What we have to learn in this Primer is—How the food becomes part and parcel of your body; how it gets oxidized; how the oxidation gives you power to move; how it is that you are able to move in all manner of ways, when you like, how you like, and as much as you like.

First of all we must learn something about the build of your body, of what parts it is made, and how the parts are put together.

7. When you want to make a snow man, you take one great roll of snow to make the body or trunk. This you rest on two thinner rolls which serve as legs. Near the top of the trunk you stick in another thin roll on either side— these you call the two arms: and lastly, on quite the top of the trunk you place a round ball for a head. Head, trunk, and limbs, *i.e.* legs and arms—these together make up a complete body.

In your snow man these are all alike, all balls of snow differing only in size and form; but in your own body, head, trunk, and limbs are quite unlike, as you might easily tell on taking them to pieces. Now you cannot very well take your own body to pieces, but you easily can that of a dead rabbit. Suppose you take one of the limbs, say a leg, to begin with.

First of all there is the **skin** with the hair on the outside. If you carefully cut this through with a knife or pair of scissors and strip it off, you will find it smooth and shiny inside. Underneath the skin you see what you call **flesh**, rather paler, not so red as the flesh of beef or mutton, but still quite like it. Covering the flesh there may be a little **fat**. In a sheep's leg as you see it at the butcher's there is a good deal of fat, in the rabbit's there is very little.

This reddish flesh you must henceforward learn to speak of as **muscle**. If you pull it about a little, you will find that you can separate it easily into parcels or slips running lengthways down the leg, each slip being fastened tight at either end, but loose between. Each slip is what is called **a muscle**. You will notice that many of these muscles are joined, sometimes at one end only, sometimes at both, to white or bluish white glistening cords or bands; made evidently of different material from the muscle itself. They are not soft and fleshy like the muscle, but firm and stiff. These are **tendons**. Sometimes they are broad and short, sometimes thin and long.

As you are separating these muscles from each other you will see (running down the leg between them) little white soft threads, very often branching out and getting too small to be seen. These are **nerves**. Between the muscles too are other little cords, red, or reddish black, and if you prick them, a drop or several drops of blood will ooze out. These are **veins**, and are not really cords or threads, but hollow tubes, filled with blood. Lying alongside the veins are similar small tubes, containing very little blood, or none at all. These are **arteries**. The **veins**and **arteries** together are called **blood-vessels**, and it will be easy for you to make out that the larger ones you see are really hollow tubes. Lastly, if you separate the muscles still more, you will come upon the hard **bone** in the middle of the leg, and if you look closely you will find that many of the muscles are fastened to this bone.

Now try to put back everything in its place, and you will find that though you have neither cut nor torn nor broken either muscle or blood-vessel or bone, you cannot get things back into their place again. Everything looks "messy." This is partly because, though you have torn neither muscle nor blood-vessel, you have torn something which binds skin and muscle and fat and blood-vessels and bone all together; and if you look again you will see that between them there is a delicate stringy substance which binds and packs them all together, just as cotton-wool is used to pack up delicate toys and instruments. This stringy packing material which you have torn and spoilt is called **connective** because it connects all the parts together.

Well, then, in the leg (and it is just the same in the arm) we have skin, fat, muscle, tendons, blood-vessels, nerves, and bone all packed together with connective and covered with skin. These together form the solid leg. We may speak of them as **the tissues** of the leg.

8. If now you turn to the trunk and cut through the skin of the belly, you will first of all see muscles again, with nerves and blood-vessels as before. But when you carefully cut through the muscles (for you cannot easily separate them from each other here), you come upon something which you did not find in

Fig. 1.—The Viscera of a Rabbit as seen upon simply opening the Cavities of the Thorax and Abdomen without any further Dissection.

A, cavity of the thorax, pleural cavity of either side; B, diaphragm; C, ventricles of the heart; D, auricles; E, pulmonary artery; F, aorta; G, lungs, collapsed, and occupying only the back part of the chest; H, lateral portions of pleural membranes; I, cartilage at the end of sternum; K, portion of the wall of body left between thorax and abdomen; a, cut ends of the ribs; L, the liver, in this case lying more to the left than the right of the body; M, the stomach; N, duodenum; O, small intestine; P, the cæcum, so largely developed in this and other herbivorous animals; Q, the large intestine.

the leg, a **great cavity**. This is something quite new— there is nothing like it in the leg—a great cavity, quite filled with something, but still a great cavity; and if you slit the rabbit right up the front of its trunk and turn down or cut away the sides as has been done in Fig. 1, you will see that the whole trunk is **hollow** from top to bottom, from the neck to the legs.

If you look carefully you will see that the cavity is divided into two by a cross partition (Fig. 1, B) called the **diaphragm**. The part **below** the diaphragm is the larger of the two, and is called the **abdomen** or belly; in it you will see a large dark red mass, which is the **liver** (L). Near the liver is the smooth pale **stomach** (M), and filling up the rest of the abdomen you will see the coils of the **intestine** or bowel, very narrow in some parts (O), very broad ($P\ Q$), broader even than the stomach, in others. If you pull the bowels on one side as you easily can do, you will find lying underneath them two small brownish red lumps, one on each side. These are the **kidneys**.

In the smaller cavity **above** the diaphragm, called the **thorax** or chest, you will see in the middle the **heart** (C), and on each side of the heart two pink bodies, which when you squeeze them feel spongy. These are the two **lungs** (G). You will notice that the heart and lungs do not fill up the cavity of the chest nearly so much as the liver, stomach, bowels, &c. fill up the cavity of the belly. In fact, in the chest there seems to be a large empty space. But as we shall see further on, the lungs did quite fill the chest before you opened it, but shrank up very much directly you cut into it, and so left the great space you see.

9. The trunk then is really a great chamber containing what are called the **viscera**, and divided into an upper and lower half, the upper half being filled with the heart and lungs, the lower with the liver, stomach, bowels, and some other organs. In front the abdomen is covered by skin and muscle only. But if all the sides of the trunk were made of such soft material it would be then a mere bag which could never keep its shape unless it were stuffed quite full. Some part of it must be strengthened and stiffened. And indeed the trunk is not a bag with soft yielding sides, but a box with walls which are in part firm and hard. You noticed that when you were cutting through the front of the chest you had to cut through several hard places. These were the **ribs** (Fig. 1, *a*), made either of hard bone or of a softer gristly substance called **cartilage**. And if you take away all the viscera from the cavity of the trunk and pass your finger along the back of the cavity, you will feel all the way down from the neck to the legs a hard part. This is the **backbone** or **vertebral column**. When you want to make a straw man stand upright you run a pole right through him to give him support. Such a support is the backbone to your own body, keeping the trunk from falling together.

In the abdomen nothing more is wanted than this backbone, the sides and front of the cavity being covered in with skin and muscle only. In the chest the sides are strengthened by the ribs, long thin hoops of bone which are fastened to the backbone behind and meet in front in a firm hard part, partly bone, partly cartilage, called the **sternum**.

But this backbone is not made of one long straight piece of bone. If it were you would never be able to bend your body. To enable you to do this it is made up of ever so many little flat round pieces of bone, laid one a-top of the other, with their flat sides carefully joined together, like so many bungs stuck together. Each of these little round flat pieces of the backbone is called a **vertebra**, and is of a very peculiar shape. Suppose you took a bung of bone, and fastened on to one side of its edge a ring of bone. That would represent a vertebra. The solid bung is what is called the **body**, and the hollow ring is what is called the **arch** of the vertebra. Now if you put a number of these bodies together one upon the top of the other, so that the bodies all came together and the rings all came together, you would have something very like the vertebral column (see Frontispiece, also Fig. 2). The bungs or bodies would make a solid jointed pillar, and the rings or arches would make together a tunnel or canal. And that is really what you have in the backbone. Only each vertebra is not exactly shaped like a bung and a ring; the body is very like a bung, but the arch is rough and jagged, and the bodies are joined together in a particular way. Still we have all the bodies of the vertebræ forming together a solid pillar which gives support to the trunk; and the arches forming together a tunnel or canal which is called the **spinal canal**, (Fig. 2, *C.S.*) the use of which we shall see

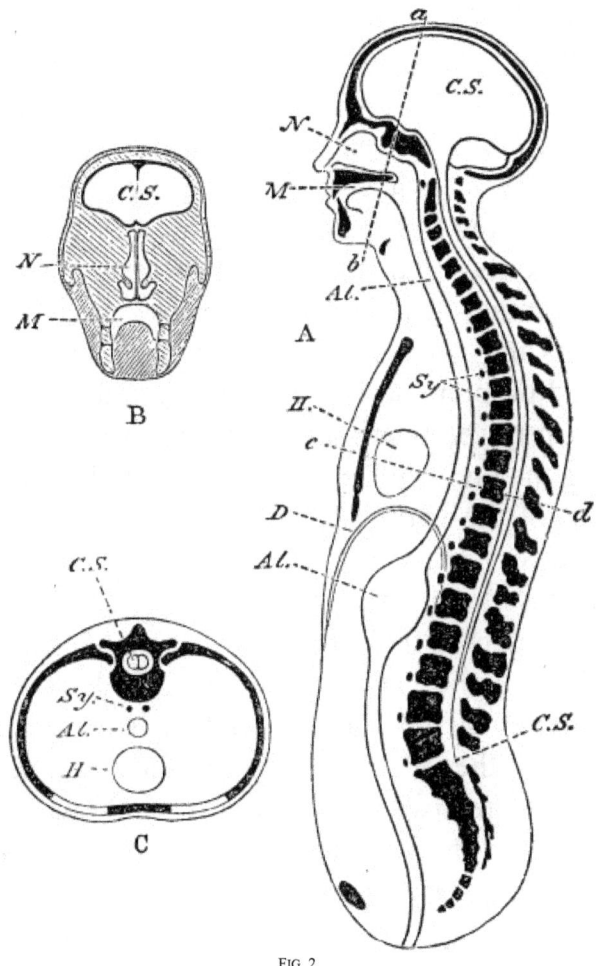

FIG. 2.

A, a diagrammatic view of the human body cut in half lengthways. C.S., the cavity of the brain and spinal cord; N, that of the nose; M, that of the mouth; Al. Al., the alimentary canal represented as a simple straight tube; H, the heart; D, the diaphragm.

B, a transverse vertical section of the head taken along the line a b; letters as before.

C, a transverse section taken along the line c d; letters as before.

directly. The round flat body of each vertebra is turned to the front towards the cavity of the trunk, and it is the row of vertebral bodies which you feel as a hard ridge when you pass your fingers down the back of the abdomen. The arches are at the back of the bodies, so you cannot feel them in the abdomen; but if you turn the rabbit on its belly and pass your finger down its back, you will feel through the skin (and you can feel the same on your own body) a sharp edge, formed by what are called the spines, *i.e.* the uneven tips of the arches of the vertebræ (Fig. 2) all the way down the back.

So that what we really have in the trunk is this. In front a large cavity, containing the viscera, and surrounded in the upper part or thorax by hoops of bone, but not (or only slightly) in the lower part or abdomen; behind, a much smaller long narrow cavity or canal formed by the arches of the vertebræ, and therefore surrounded by bone all the way along, and containing we shall presently see what; and between these two cavities, separating the one from the other, a solid pillar formed by the bodies of the vertebræ. So that if you were to take a cross slice, or transverse section as it is called, of the rabbit across the chest, you would get something like what is represented in Fig. 2, C, where *C.S.* is the narrow canal of the arches and where the broad cavity of the chest containing the heart *H* is enclosed in the ribs reaching from the vertebra behind to the sternum in front. Both cavities are covered up on the outside with muscles, blood-vessels, nerves, connective, and skin, just as in the leg.

10. We have now to consider the **head and neck**. If you cut through the skin of the neck of the rabbit, you will see, first of all, muscles and nerves, and several large blood-vessels; but you will find no large cavity like that in the trunk. So far the neck is just like the leg. But if you look carefully you will see two tubes which are not blood-vessels, and the like of which you saw nowhere in the leg. One of these tubes is firm, with hardish rings in it; it is the windpipe or **trachea**; the other is soft, and its sides fall flat together; this is the gullet or **œsophagus**, leading from the mouth to the stomach. Behind these and the muscles in which they run you will find, just as in the trunk, a vertebral column, without ribs, but composed of bodies, and behind the bodies there is a vertebral canal. This vertebral column and vertebral canal in the neck are simply continuations of the vertebral column and canal of the trunk.

The neck, then, differs from the leg in having a vertebral column and canal with a trachea and œsophagus, and differs from the trunk in having no cavity and no ribs.

The head, again, is unlike all these. Indeed, you will not understand how the head is made unless you take a rabbit's skull and place it side by side with the rabbit's head. If you do this, you will at once see how the mouth and throat are formed. **You will notice that the skull is all in one piece**, except a bone which you will at once recognize as the jawbone, or, to speak more correctly, the lower **jawbone**; for there are two jawbones. Both these carry teeth, but the upper one is simply part of the skull, and does not move; the lower one does move; it can be made to shut close on the upper jaw, or can be separated a good way from it. The opening between the two jaws is the gap or gape of the mouth, which as you know can be opened or shut at pleasure. If you try it on yourself you will find that, as in the rabbit, it is the lower jaw which moves when you open or shut your mouth. The upper jaw does not move at all except when your whole head moves. Underneath the skull at the top of the neck the mouth narrows into the throat, into the upper part of which the cavity of the nose opens. So that there are two ways into the throat, one through the mouth and the other through the nose (Fig. 2).

At the back of the skull you will see a rounded opening, and if you put a bodkin through this opening you will find it leads into a large hollow space in the inside of the skull. In the living rabbit this hollow space is filled up with the **brain**. The skull, in fact, is a box of bone to hold the brain, a bony brain-case. This bony case fits on to the top of the vertebræ of the neck in such a way that the rounded opening we spoke of just now is placed exactly over the top of the tunnel or canal formed by the rings or arches of the vertebræ. If you were to put a wire through the arch of the lowest vertebra, you might push it up through the canal formed by the arches of all the vertebræ, right into the brain cavity. In fact the brain-case and the row of arches of the vertebræ form together one canal, which is a narrow tube in the back and in the neck, but swells out in the head into a wide rounded space (Fig. 2, A and B, *C.S.*) During life this canal is filled with a peculiar white delicate material, which is called **nervous matter**. The rounded mass of this material which fills up the cavity of the skull is called the **brain**; the narrower, rod-like, or band-like mass which runs down the vertebral canal in the neck and back is called the **spinal cord**. They have separate names, but they are quite joined together, and the rounded brain tapers off into the band-like cord in such a way that it is difficult to say where the one begins and the other ends.

11. In the skull, besides the larger openings we have spoken of, you will find several small holes leading from the outside of the skull into the inside of the brain-case. Some of these holes are filled up during life by blood-vessels, but in others run those delicate white threads or cords which you have already learnt to call nerves. **Nerves are in fact branches of nervous material running out from the brain or spinal cord.** Those from the brain pass through holes in the skull, and at first sight seem to spread out very irregularly. Those which branch off from the spinal cord are far more regular. A nerve runs out on each side between every two vertebræ, little rounded gaps being left for that purpose where the vertebræ fit together, so that when you look at a spinal cord with portions of the nerves still connected with it, it seems not unlike a double comb with a row of teeth on either side. The nerves which spring in this way from the spinal cord are called **spinal nerves**, and soon after they leave the vertebral canal they divide into branches, and so are spread nearly all over the body. In any piece of skin or flesh you examine, never mind in what part of the body, you will find nerves and blood-vessels. If you trace the nerves out in one direction, you will find them joining together to form larger nerves, and these again joining others, till at last all end in either the spinal cord or the brain. If you try to trace the same nerves in the other direction, you will find them branching into smaller and smaller nerves, until they become too small to be seen. If you take a microscope you will find they get still smaller and smaller until they become the very finest possible threads.

The blood-vessels in a similar way join together into larger and larger tubes, which last all end, as we shall see, in the heart. **Every part of the body, with some few exceptions, is crowded with nerves and blood-vessels. The nerves all come from the brain or spinal cord— the vessels from the heart. So that every part of the body is governed by two centres, the heart, and the brain or spinal cord.** You will see how important it is to remember this when we get on a little further in our studies.

12. Well, then, the body is made up in this way. First there is the head. In this is the skull covered with skin and flesh, and containing the brain. The skull rests on the top of the backbone, where the head joins the neck. In the upper part of the neck, the throat divides into two pipes or tubes—one the windpipe, the other the gullet. These running down the neck in front of the vertebral column, covered up by many muscles, when they get about as far down as the level of the shoulders, pass into the great cavity of the body, and first into the upper part of it, or chest.

Here the windpipe ends in the lungs, but the gullet runs straight through the chest, lying close at the back on the backbone, and passes through a hole in the diaphragm into the abdomen, where it swells out into the stomach. Then it narrows again into the intestine, and after winding about inside the cavity of the abdomen a good deal, finally leaves it.

You see the **alimentary canal** (for that is the name given to this long tube made up of gullet, stomach, intestine, &c.) **goes right through the cavity of the body without opening into it**—very much as the tall narrow glass of a lamp passes through the large globe glass. You might pour anything down the narrow glass without its going into the globe glass, and you might fill the globe glass and yet leave the narrow glass quite empty. If you imagine both glasses soft and flexible instead of hard and stiff, and suppose the narrow glass to be very long and twisted about so as to all but fill the globe, you will have a very fair idea of how the alimentary canal is placed in the cavity of the body.

Besides the alimentary canal, there is in the chest, in addition to the windpipe and lungs, the heart with its great tubes, and in the abdomen there are the liver, the kidneys, and other organs.

These two great cavities, with all that is inside them, together with wrappings of flesh and skin which make up the walls of the cavities, form the trunk, and on to the trunk are fastened the jointed legs and arms. These have no large cavities, and the alimentary canal goes nowhere near them.

One more thing you have to note. There is only one alimentary canal, one liver, one heart—but there are two kidneys and two lungs, the one on one side, the other on the other, and the one very much like the other. There are two arms and two legs, the one almost exactly like the other. There is only one head, but one side of the head is almost exactly like the other. One side of the vertebral column is exactly like the other—as are also the two halves of the brain and the two halves of the spinal cord.

In fact, if you were to cut your rabbit in half from his nose to his tail, you would find that except for his alimentary canal, his heart, and his liver, one half was almost exactly the counterpart of the other.

Such is the structure of a rabbit, and your own body, in all the points I have mentioned, is made up exactly in the same way.

13. Let us now go back to the question. **How is it that we can move about as we do?** And first of all let us take one particular movement and see if we can understand that.

For instance, you can bend your arm. You know that when your arm is lying flat on the table, you can, if you like, bend the lower part of your arm (the fore-arm as it is called, reaching from the elbow to the hand) on the upper arm until your fingers touch your shoulder. How do you manage to do that?

Look at the bones of the arm in a skeleton. (Frontispiece; also Fig. 3.) You will see that in the upper arm there is one rather large bone (*H*) reaching from the shoulder to the elbow, while in the fore-arm there are two, one (*U*) being wider and stouter than the other (*Ra*) at the elbow, but smaller and more slender at the wrist. The bone in the upper arm is called the **humerus**; the bone in the fore-arm, which is stoutest at the elbow, is called the **ulna**; the one which is stoutest at the wrist is called the **radius**. If you look carefully you will see that the end of the humerus at the elbow is curiously rounded, and the end of the ulna at the elbow curiously scooped out, in such a way that the one fits loosely into the other.

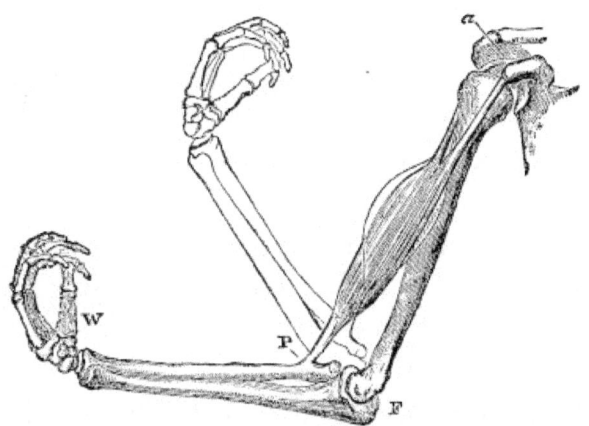

Fig 3.—The Bones of the Upper Extremity with the Biceps Muscle.

The two tendons by which this muscle is attached to the scapula, or shoulder-blade, are seen at a. P indicates the attachment of the muscle to the radius, and hence the point of action of the power; F, the fulcrum, the lower end of the humerus on which the upper end of the radius (together with the ulna) moves; W, the weight (of the hand).

If you try to move them about one on the other, you will find that you can easily double the ulna very closely on the humerus without their ends coming apart, and if you notice you will see that as you move the ulna up and down, its end and the end of the humerus slide over each other. But they will only slide one way, what we may call up and down. If you try to slide them from side to side, you will find that they get locked. They have only one movement, like that of a door on its hinge, and that movement is of such a kind as to double the ulna on the humerus.

Moreover, if you look a little more carefully you will find that, though you can easily double the ulna on the front of the humerus, and then pull it back again until the two are in a straight line, you cannot bend the ulna on the back of the humerus. On examining the end of the ulna you will find at the back of it a beak-like projection (Fig. 3, also Frontispiece), which when the bones are straightened out locks into the end of the humerus, and so prevents the ulna being bent any further back. This is the reason why you can only bend your arm one way. As you very well know, you can bend your arm so as to touch the top of your shoulder with your fingers, but you can't bend it the other way so as to touch the back of your shoulder; you can't bring it any further back than the straight line.

14. Well, then, at the elbow the two bones, the humerus and ulna, are so shaped and so fit into each other that the arm may be straightened or bent. In the skeleton the two bones are quite separate, *i.e.* they have to be fastened together by something, else they would fall apart. Most probably in the skeleton you have been examining they are fastened together by wires or slips of brass. But they would hold together if you took away the wire or brass slips and bound some tape round the two ends, tight enough to keep them touching each other, but loose enough to allow them to move on each other. You might easily manage it if you took short slips of tape, or, better still, of india-rubber, and placed them all round the elbow, back, front, and sides, fastening one end of each slip to the humerus and the other to the ulna. If you did this you would be imitating very closely the manner in which the bones at the elbow are kept together in your own arm. Only the slips are not made of india-rubber, but are flat bands of that stringy, or as we may now call it fibrous stuff, which in the preceding lessons you learnt to call connective tissue. These flat bands have a special name, and are called **ligaments**.

At the elbow the two ends of the ulna and humerus are kept in place by ligaments or flat bands of connective tissue.

In the skeleton, the surfaces of the two bones at the elbow where they rub against each other, though somewhat smooth, are dry. If you ever looked at the knuckle of a leg of mutton before it was cooked, you will have noticed that you have there two bones slipping over each other somewhat as they do at the elbow, and will remember that where the bones meet they are wonderfully smooth, and very moist, so as to be quite slippery. It is just the same in your own elbow; the end of the ulna and the end of the humerus are beautifully smooth and quite moist, so that they slip over each other as easily as possible. You know that your eye is always moist. It is kept moist by tears, though you don't speak of tears until your eyes overflow with moisture; but in reality you are always crying a little. Well, there are, so to speak, tears always being shed inside the wrapping of ligaments around the elbow, and they keep the two surfaces of the bones continually moist.

The ends of bones where they touch each other are also smooth, because they are coated over with what is called gristle or cartilage. Bone is very hard and very solid; there is not much water in it. Bones dry up very little. Cartilage is not nearly so hard as bone; there is very much more water in it. When it is quite fresh it is very smooth, but because it has a good deal of water in it, it shrinks very much when it dries up, and when dried is not nearly so smooth as when it is fresh. You can see the dried-up cartilage on the ends of the bones in the skeleton—it is somewhat smooth still, but you can form no idea of how smooth it is in the living body by simply seeing it on the dried skeleton.

At the elbow, then, we have the ends of two bones fitting into each other, so that they will move in a certain direction; these ends are smoothed with cartilage, kept moist with a fluid, and held in place by ligaments. All this is a called a joint.

15. There are a great many other joints in the body besides the elbow-joint: there is the shoulder-joint, the knee-joint, the hip-joint, and so on. These differ from the elbow-joint in the shape of the ends of the bone, in the way the bones move on each other, and in several other particulars, but we must not go into these differences now. They are all like the elbow, since in each case one bone fits into another, the surfaces are coated with cartilage, are kept moist with fluid (what the grooms call joint-oil, though it is not an oil at all), and are held in place by ligaments.

I dare say you will have noticed that though I have been speaking only of the humerus and ulna at the elbow, the other bone of the fore-arm—the radius—has something to do with the elbow too. I left it out in order to simplify matters, but it is nevertheless quite true that the end of the humerus moves over the end of the radius as well as over the end of the ulna, and that the end of the radius is also coated with cartilage and is included in the wrapping of the ligaments. I might add that the radius also moves independently on the ulna, but I don't want to trouble you with this just now. What I wanted to show you was that the elbow is a joint, a joint so constructed that it allows the fore-arm to be bent on the upper arm.

16. In order that the arm may be bent, some force must be used. The ulna or radius—for the two move together—must be pushed or pulled towards the humerus, or the humerus must be pushed or pulled towards the radius and ulna. How is this done in your own arm?

Take the bones of the arm; fix the top end of the humerus; tie it to something so that it cannot move. Fasten a piece of string to either the radius or ulna (it doesn't matter which), rather near the elbow. Bore a hole through the top of the humerus and pass the string through it. Your string must be long enough to let the arm be quite straight without any strain on the string. Now, taking hold of the string where it comes out through the humerus, pull it. The fore-arm will be bent on the arm. Why? **Because you have been working a lever of the third order.**

The radius and ulna form the lever; its fulcrum is the end of the humerus in the elbow (Fig. 3, F); the weight to be moved is the weight of the radius and ulna (with that of the bones of the hand if present), and this may be represented by a weight applied at about the middle of the fore-arm; the power is the pull you give the string, and that is brought to bear on your lever at the point where the string is fastened to the radius, *i.e.* nearer the fulcrum than the point where the weight is applied; and you know that when you have the fulcrum at one end and the power between the fulcrum and the weight, you have a lever of the third order.

Now, in order to make the thing a little more like what takes place in your own arm, instead of boring a hole through the humerus, let the string glide in a groove which you will see at the top of the humerus, and fasten the end of it to the shoulder-blade or anything you like above the humerus, and let the string be just long enough to let the arm be quite straightened out, but no longer, so that when the arm is straight the string is just about tight, or at least not loose.

Now shorten the string by pinching it up into a loop. Whenever you do this you will bend the fore-arm on the arm. Suppose you used a string which you had not to pinch up, but which, when you pleased, you could make to shorten itself. **Every time it shortened itself it would pull the fore-arm up and would bend the arm—and every time it slackened again, the arm would fall back into the straight position.**

In your arm there is not a string, but a body, placed very much as our string is placed, and which has the power of shortening itself when required. Every time it shortens itself it bends the arm, and when it has done shortening and lengthens again, the arm falls back into its straight position. This body which thus can shorten and lengthen itself is called a muscle.

If you put one hand on the front of your other upper arm, about half-way between your shoulder and elbow, and then bend that arm, you will feel something rising up under your hand. This is the muscle, which bends the arm, shortening, or, as we shall learn to call it, **contracting**.

In your own arm, as in the limb of the rabbit which you studied in your last lesson, the flesh is arranged in masses or bundles of various sizes and shapes, and each mass or bundle is called a muscle. There are several muscles in the arm, but there is in particular a large one occupying the front of the arm, called the **biceps**. It is a rounded mass of red flesh, considerably longer than it is broad or thick, and tapering away at either end. It is represented in Fig. 3.

You may remember that while examining the leg of the rabbit you noticed that in many of the muscles, the soft flesh, which made up the greater part of the muscle, at one or both ends of the muscle suddenly left off, and changed into much firmer material which was white and glistening. This firmer white part you were told was called the tendon of the muscle. The rest of the muscle, generally called "the belly," is made up of what you are accustomed to call flesh, or lean meat, but which you must now learn to speak of as **muscular substance**. Every muscle, in fact, consists in the first place of a mass of muscular substance. **This muscular substance is**

made up of an immense number of soft strings or fibres, all running in one direction and done up into large and small bundles. At either end of the muscle these soft muscular fibres are joined on to firmer but thinner fibres of connective or fibrous tissue. And these thinner but firmer fibres make up the cord or band of tendon with which the muscle finishes off at either end.

It is by these tendons that the soft muscles are joined on to the hard bones, or to some of the other firm textures of the body. The tendons are sometimes round and cord-like, sometimes flat and spread out. Sometimes they are very long, sometimes very short, so as to be scarcely visible. But always you have some amount of the firmer fibres of connective tissue joining the soft muscular fibres on to the bones, and generally the tendons are not only firmer but much thinner and more slender than the belly of the muscle.

The muscular belly of the biceps is placed in the front of the upper arm. Some little way above the elbow-joint it ends in a small round strong tendon which slips over the front of the elbow and is fastened to, *i.e.* grows on to, the radius at some little distance below the joint (Fig. 3, P). The upper part of the muscular belly ends a little below the shoulder, not in one tendon but in two[1] tendons (Fig. 3, *a*), which gliding over the end of the humerus are fastened to the shoulder-blade (or **scapula** as it is called), into which the humerus fits with a joint.

We have then in the biceps a thick fleshy muscular belly placed in the front of the arm and fastened by tendons, at one end to the shoulder-blade, and at the other to the fore-arm. What would happen if when the arm is straight and the shoulder-blade fixed, the biceps were suddenly to grow very much shorter than it was? Evidently the same thing that happened when you pinched up and shortened the string which, if you look back you will see, we supposed to be placed very much as the biceps with its tendons is placed. **The radius and ulna would be pulled up, the fore-arm would be bent on the arm.**

Now tendons have no power of shortening themselves, but muscular substance does possess this remarkable power of suddenly shortening itself. Under certain circumstances each soft muscular fibre of which the muscle is made will suddenly become shorter, and thus the whole muscle becomes shorter, and so pulls its two tendinous ends closer together, and if one end be fastened to something fixed, and the other to something moveable, the moveable thing will be moved.

This way that a muscle or a muscular fibre has of suddenly shortening itself is called a **muscular contraction. All muscles, all muscular fibres, have the power of contracting.** Now a mass of substance like the biceps might grow shorter in two ways. It might squeeze itself together and become smaller altogether, it might squeeze itself as you would squeeze a sponge into a smaller bulk. Or it might change its form and not its bulk, becoming thicker as it became shorter, just as you might by pressing the two ends together squeeze a long thin roll of soft wax into a short thick one. It might get shorter in either of these two ways, but it does actually do so in the latter way; it gets thicker at the same time that it gets shorter, and gets nearly as much thicker as it gets shorter. And that is why, when you put your hand on the arm which is being bent, you feel something rise up. **You feel the biceps getting thicker as it is getting shorter in order to bend the arm.**

The shortening does not last for ever. Sooner or later the muscle lengthens again, getting thinner once more, and so returns to its former state. The lengthened condition of the muscle is the natural condition, the condition of rest. The shortening or contraction is an effort which can only be continued for a certain time. The contraction bends the arm, and as long as the muscle remains shortened the arm keeps bent; but as the muscle lengthens, the weight of the hand and fore-arm, if there is nothing to prevent, straightens the arm out again.

It is in the muscle alone, in the belly made up of muscular fibres, that the shortening takes place. The tendons do not shorten at all. On the contrary, if anything they lengthen a little, but only a very little, when the muscle pulls upon them. Their purpose is to convey to the bone the pull of the muscle. They are not necessary, only convenient. It would be possible but awkward to do without them. Suppose the fleshy fibres of the biceps reached from the shoulder-blade to the fore-arm: you could bend your arm as before, but it would be very tiresome to have the muscle swelling up in the inside of the elbow, or on the top of the shoulder; in either place it would be very much in the way. By keeping the fleshy, the real contracting muscle, in the arm, and carrying the thin tendons to the arm and to the shoulder, you are enabled to do the work much more easily and conveniently.

Well, then, we have got thus far in understanding how the arm is bent. The biceps muscle contracts and shortens, tries to bring its two tendinous ends together. The upper tendons, being fastened to the fixed shoulder-blade, cannot move; but the lower tendon is fixed to the radius; the radius, with the ulna to which it is fastened, readily moves up and down on the elbow-joint—the shape of bones in the joint and all the arrangements of the joint, as we have seen, readily permitting this. When the muscle, then, pulls on its lower tendon, its pulls on the radius at the point where the tendon is fastened on to the bone. The radius thus pulled on forms with the ulna a lever of the third order, working on the end of the humerus as a fulcrum; and thus as the tendon is pulled the fore-arm is bent.

17. But now comes the question. What makes the muscle shorten or contract? You willed to move your arm, and moved it, as we have seen, by making the biceps contract; but **how did your will make the biceps contract**?

If you could examine your arm as you did the leg of the rabbit, you would find running into your biceps muscle, one or more of those soft white threads or cords, which you have already learnt to recognize as **nerves**.

These nerves seem to grow into and be lost in the biceps muscle. We need not follow them any further in that direction, but if we were to trace them in the other direction, up the arm, we should find that they soon meet with other similar nerves, and that the several nerves joining together form stouter and thicker nerve-cords. These again join others, and so we should proceed until we came to quite stoutish white nerve-trunks as they are called, which we should find passed at last between the vertebræ, somewhere in the neck, into the inside of the vertebral canal, where they became mixed up with the mass of nervous material we have already spoken of as the spinal cord.

What have these nerves to do with the bending of the arm? Why simply this. Suppose you were able without much trouble to cut across the delicate nerves going to your biceps, and did so: what would happen? You would find that you had lost all power of bending your arm; however much you willed it, there would be no swelling rise up in your arm. Your biceps would remain perfectly quiet, and would not shorten at all, would not contract in obedience to your will.

What does this show? It proves that when you will to bend your arm, something passes along the nerves going to the biceps muscle, which something causes that muscle to contract? The nerve, then, is a bridge between your will and the muscle—so that when the bridge is broken or cut away, the will cannot get to the muscle.

If anywhere between the muscle and the spinal cord you cut the nerve which goes, or branches from which go, to the muscle, you destroy the communication between the will and the muscle.

The spinal cord, as we have seen, is a mass of nervous substance continuous with the brain; from the spinal cord nearly all the nerves of the body are given off; those nerves whose branches go to the biceps muscle in the arm leave the spinal cord somewhere in the neck.

If you had the misfortune to have your spinal cord cut across or injured in your neck, you might still live, but you would be paralysed. You might will to bend the arm, but you could not do it. You would know you were willing, you would feel you were making an effort, but the effort would be unavailing. The spinal cord is part of the bridge between the will and the muscle.

When you bend your arm, then, this is what takes place. **By the exercise of your will a something is started in your brain. That something—we will not stop now to ask what that something is—passes from your brain to the spinal cord, leaves the spinal cord and travels along certain nerves, picking its way among the intricate bundles of delicate nervous threads which run from the upper part of the spinal cord to the arm until it reaches the biceps muscle. The** muscle, directly that "something" comes to it along its nerves, contracts, shortens, and grows thick; it rises up in the arm; its lower tendon pulls at the radius; the radius with the ulna moves on the fulcrum of the humerus at the elbow-joint, and the arm is bent.

You wish to leave off bending the arm. Your will ceases to act. The something to which your will had given rise dies away in the brain, dies away in the spinal cord, dies away in the nerves, even in the finest twigs. The muscle, no longer excited by that something, ceases to contract, ceases to swell up, ceases to pull at the radius, and the fore-arm by its own weight falls into its former straightness, stretching, as it falls, the muscle to its natural length.

18. So far I hope you have followed me, but we are still very far from being at the bottom of the matter. Why does the muscle contract when that something reaches it through the nerves? We must content ourselves by saying that it is the property of the muscle to do so. Does the muscle always possess this property? No, not always.

Suppose you were to tie a cord very tightly round the top of your arm, close to the shoulder. What would happen? If you tied it tight enough (I don't ask you to do it, for you might hurt yourself) the arm would become pale, and very soon would begin to grow cold. It would get numbed, and would gradually seem to grow very heavy and clumsy; your feeling in it would be blunted, and after a while be altogether lost. When you tried to bend your arm you would find great difficulty in doing so. Though you tried ever so much, you could not easily make the biceps contract, and at last you would not be able to do so at all. You would discover that you had lost all power of bending your arm. And then if you undid the cord you would find that after some very uncomfortable sensations, little by little the power would come back to you; the arm would grow warm again, the heaviness and clumsiness would pass away, the feeling in it would return, you would be able to bend it, and at last all would be as it was before.

What did you do when you tied the cord tight? The chief thing you did was to press on the blood-vessels in the arm and so stop the blood from moving in them. If instead of tying the cord round the whole arm you had tied a finer thread round the blood-vessels only, you would have brought about very nearly the same effect. We saw in the last lesson how all parts of the body are supplied with blood-vessels, with veins, and arteries. In the arm there is a very large artery, branches from which go all over the arm. Some of these branches go to the biceps muscle. What would happen if you tied these branches only, tying them so tight as to stop all the blood in them, but not interfering with the blood-vessels in the rest of your arm? The arm as a whole would grow neither pale nor cold, it would not become clumsy or heavy, you would not lose your feeling in it, but nevertheless if you tried to bend your arm you would find you could not do it. You could not make the biceps contract, though all the rest of the arm might seem to be quite right.

What does this teach us? **It teaches us that the power which a muscle has of contracting when called upon to do so, may be lost and regained, and that it is lost when the blood is prevented from getting to it.** When a cord is tied round the whole arm, the power of the whole arm is lost. This loss of power is the beginning of death, and indeed if the cord were not unloosed the arm would quite die—would mortify, as it is said. When only those blood-vessels which go to the biceps are tied, the biceps alone begins to die, all the rest of the arm remaining alive, and the first sign of death in the biceps is the loss of the power to contract when called upon to do so.

In order that you may bend your arm, then, you must not only have a biceps muscle with its nerves, its tendons, and all its arrangements of bones and joints, but the muscle must be supplied with blood.

19. We can now go a step further and ask the question, **What is there in the blood that thus gives to the muscle the power of contracting, that in other words keeps the muscle alive?** The answer is very easily found. What is the name commonly given to this power of a muscle to contract? We generally call it strength. Lay your arm straight out on the table, put a heavy weight in your hand, and try to bend your arm. If you could do it, one would say you were strong; if you could not, one would say you were weak—all the stronger or weaker, the heavier or lighter the weight. In the one case your biceps had great power of contracting; in the other, little power. Try and find out the heaviest weight you can raise in this way by bending your arm, some morning, not too long after breakfast, when you are fresh and in good condition. Go without any dinner, and in the afternoon or evening, when you are tired and hungry, try to raise the same weight in the same way. You will not be able to do it. Your biceps will have lost some of its power of contracting, will be weaker than it was in the morning. What makes it weak? The want of food. But how can the food affect the muscle? You do not place the food in the muscle; you put it into your mouth, and from thence it goes into your stomach and into the rest of your alimentary canal, and there seems to disappear. How does the food get at the muscle? By means of the blood. The food becomes blood. **The things which you eat as food**

become changed into other things which form part of the blood. Those things going to the muscle give it strength and enable it to contract. And that is why food makes you strong.

20. But you are always wanting food day by day, from time to time. Why is that? Because the muscle in getting strength out of the food changes it, uses it up, and so is always wanting fresh blood and new food. We have seen in Art. 1 that food is fuel. We have also seen that muscle (and other parts of the body do the same) is always burning, burning without flame but with heat, burning slowly but burning all the same, and doing the more work the more it burns. The fuel it burns is not dry wood or coal, but wet, watery blood, a special kind of fuel prepared for its private use, in the workshop of the stomach or elsewhere, out of the food eaten by the mouth. This it is always using up; of this it must always have a proper supply, if it is to go on working. Hence there must always be fresh blood preparing; hence there must from time to time be fresh supplies of food out of which to manufacture fresh blood.

To understand then fully what happens when you bend your arm, we have to learn not only what we have learnt about the bones and the joint and the muscle and the nerves, about the machinery and the engine, we have to study also how the food is changed into blood, how the blood is brought to the muscle, what it is in the blood on which the muscle lives, what it is which the muscle burns, and how the things which result from the burning, the ashes and the smoke or carbonic acid and the rest of them, are carried away from the muscle and out of the body.

Meanwhile let me remind you that for the sake of being simple I have been all this while speaking of one muscle only, the biceps in the arm. But there are a multitude of muscles in the body besides the biceps, as there are many bones besides those of the arm, and many joints besides the elbow. But what I have said of the one is in a general way true of all the rest. The muscles have various forms, they pull upon the bones in various ways, they work on levers of various kinds. The joints differ much in the way in which they work. All manner of movements are produced by muscles pulling sometimes with and sometimes against each other. But you will find when you come to examine them that all the movements of which your body is capable depend at bottom on this— **that certain muscular fibres, in obedience to a something reaching them through their nerves, contract, shorten, and grow thick, and so pull their one end towards the other, and that to do this they must be continually supplied with pure blood.**

Moreover, what I have said of the relations of muscle to blood is also true of all other parts of the body. Just as the muscle cannot work without a due supply of blood, so also the brain and the spinal cord and the nerves have even a more pressing need of pure blood. The weakness and faintness which we feel from want of food is quite as much a weakness of the brain and of the nerves as of the muscles,—perhaps rather more so. And other parts of the body of which we shall have to speak later on need blood too.

The whole history of our daily life is shortly this. The food we eat becomes blood, the blood is carried all over the body, round and round in the torrent of the circulation; as it sweeps past them, or rather through them, the muscle, the brain, the nerve, the skin pick out new food for their work and give back the things they have used or no longer want. As they all have different works, some use up what others have thrown away. There are, besides, scavengers and cleaners to pick up things no longer wanted anywhere and to throw them out of the body. Thus the blood is kept pure as well as fresh. Through the blood thus ever brought to them, each part does its work: the muscle contracts, the brain feels and wills, the nerves carry the feeling and the willing, and the other organs of the body do their work too, and thus the whole body is kept alive and well.

21. What, then, is this blood which does so much?

Did you ever look through a good microscope at the thin transparent web of a frog's foot, and watch the red blood coursing along its narrow channels? If not, go and look at it at once; you will never understand any physiology till you have done so. There you will see a network of delicate passages far finer than any of your own hairs, and through those passages a tumbling crowd of tiny oval yellow globules hurrying and jostling along. Some of the passages are wider than others, and through some of the wider ones you will see a thick stream of globules rushing onwards towards the smaller channels, and spreading out among them. The globules which you see are floating in a fluid so clear that you cannot see it. Some of the smaller channels are so narrow that only one globule or **corpuscle**, as we may call it, can pass through at a time, and very frequently you may see them passing in single file. Watching them as they glide along these narrow paths, you will note that at last they tumble again into wider passages, somewhat like those from which they came, except that the stream runs away from instead of towards the narrower channels; and in the stream the corpuscle you are watching shoots out of sight. The finest passages are called **capillaries**; they are guarded by delicate walls which you can hardly see; they seem to you passages only, and how fine and small they are will come home to you when you recollect that all you are looking at is going on in the depths of a skin which is so thin that perhaps you would be inclined to say it has no thickness at all.

The larger channels which are bringing the blood down to the capillaries are the ends of vessels like those which in the rabbit you learnt to call arteries, and the other larger channels through which the blood is rushing away from the capillaries are the beginnings of veins.

When you have watched this frog's foot for some little time, turn away and reflect that in almost every part of your own body, in every square inch, in almost every square line, something very similar might be seen could the microscope be brought to bear upon it, only the corpuscles are smaller and round, the capillaries narrower and for the most part more thick-set, and the race a swifter one. In the muscle of which we were speaking in the last lesson, each of the soft long fibres of which the muscle is composed is wrapped round with a close network of these tiny capillaries, through which, as long as life lasts, for ever rushes a swift stream of blood, reddened by countless numbers of tiny corpuscles.

In every part of your flesh, in your brain and spinal cord, in your skin, your bones, your lungs, in all organs and in nearly every part of your body, there is the same hurrying rush through narrow tubes of red corpuscles and of the clear fluid in which these swim.

If you prick your finger it bleeds. Almost any part of your body would bleed were you to prick it. So thick-set are the little blood-vessels, that wherever you thrust a needle, be it as fine a needle as you please, you will be sure to pierce and tear some little blood channel, either artery or capillary or vein, and out will come the ruddy drop.

22. What is blood? It is a fluid; it runs about like water: yet it is thicker than water, thicker for two reasons. In the first place, water, that is pure water, is all one substance. If you were to look at it with ever so powerful a microscope, you would see nothing in it. It is exceedingly transparent—you can see very well through ever such a thickness of clean water. But if you were to try and look through even a very thin sheet of blood spread out between two glass plates, you would find that you could see very little; **blood is very opaque.** If again you examine a drop of your blood with a microscope, what do you see? **A number of little**

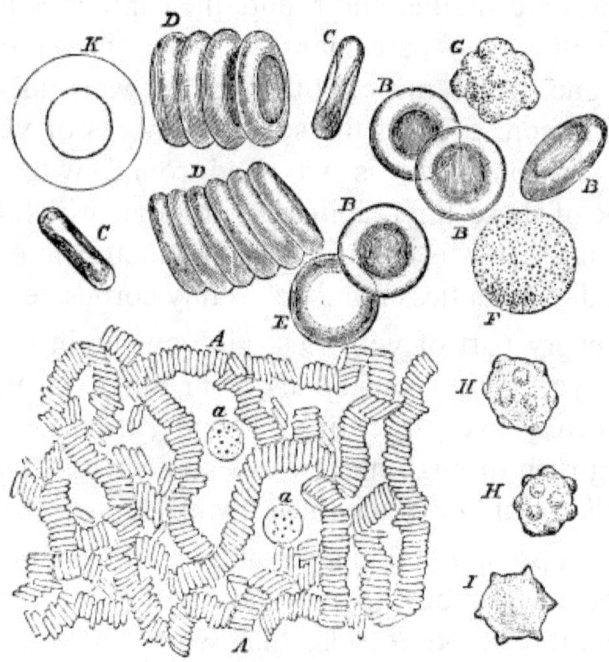

Fig. 4.—Red and White Corpuscles of the Blood magnified.

A. Moderately magnified. The red corpuscles are seen lying in rows like rolls of coins; at a and a are seen two white corpuscles.

B. Red corpuscles much more highly magnified, seen in face; C. ditto, seen in profile; D. ditto, in rows, rather more highly magnified; E. a red corpuscle swollen into a sphere by imbibition of water.

F. A white corpuscle magnified same as B.; G. ditto, throwing out some blunt processes; K. ditto, treated with acetic acid, and showing nucleus, magnified same as D.

H. Red corpuscles puckered or crenate all over.

I. Ditto, at the edge only.

round bodies, the blood discs or blood corpuscles (Fig. 4, *A*). If you look carefully you will notice that most of them are round, as *B*; but every now and then you see something like *C*. That is one of the round ones seen sideways; for they are not round or spherical like a ball, but circular and dimpled in the middle, something like certain kinds of biscuit. When you see one by itself it looks a little yellow in colour, that is all; but when you see them in a lump, the lump is clearly red. Remember how small they are: three thousand of them put flat in a line, edge to edge, like a row of draughts, would just about stretch across one inch. All the redness there is in blood belongs to them. When you see one of them, you see so little of the redness that it seems yellow. If you were to put a drop of blood into a tumbler of water, the water would not be stained red, but only just turned of a yellowish tint, so little redness would be given to it by the drop of blood. In the same way a very very thin slice of currant jelly would look yellowish, not red.

These red corpuscles are not hard solid things, but delicate and soft, very tender, very easily broken to pieces, more like the tiniest lumps of red jelly than anything else, and yet made so as to bear all the squeezing which they get as they are driven round and round the body.

Besides these red corpuscles, you may see if you look attentively **other little bodies, just a little bigger than the red corpuscles, not coloured at all, and not circular and flat, but quite round like a ball** (Fig. 4, *a*, *F*, *G*). That is to say, these are very often quite round, only they have a curious trick of changing their form. Imagine you were looking at a suet dumpling so small that about two thousand five hundred of them could be placed side by side in the length of one inch—and suppose the round dumpling while you were looking at it gradually changed into the shape of a three-cornered tart, and then into a rounded square, and then into the shape of a pear, and then into a thing that had no shape at all, and then back again into a round ball, and kept doing this apparently all of its own accord while you were looking at it—wouldn't you think it very curious? Well, one of these little bodies in the blood of which we are speaking, and which are called white corpuscles, may be seen, when a drop of blood is watched under the microscope, to go on in this way, continually changing its shape. But of these **white corpuscles** of the blood, and of their wonderful movements, you will learn more as you go on in your physiological studies.

23. Besides these red and white corpuscles there is nothing else very important in the blood that you can *see* with the microscope; but their being in the blood is one reason why blood is thicker than water.

Did you ever see a pig or sheep killed? If so, you would be sure to notice that the blood ran quite fluid from the blood-vessels in the neck, ran and was spilt like so much water—but that very soon the blood caught in the pail or spilt on the stones became quite solid, so that you could pick it up in lumps. Whenever blood is shed from the living body, within a short time it becomes solid. This becoming solid is called the **clotting** or **coagulation of blood**.

What makes it clot? Suppose while the blood was running from the pig's neck into the butcher's pail, and while it was still quite fluid, you were to take a bunch of twigs and keep slowly stirring the blood round and round in the pail. You would naturally expect that the blood would soon begin to clot, would get thicker and thicker and more and more difficult to stir. But it does not; and if you keep on stirring long enough you will find that it never clots at all. **By continually stirring it you will prevent its clotting.** Now take out your bundle of twigs: you will find it covered all over with a thick reddish mass of some soft sticky substance; and if you pump on the red mass you will be able to wash away all its red colour, and will have nothing left but a quantity of white, soft, sticky, stringy material, all entangled and matted together among the twigs of your bundle. This stringy material is in reality made up of a number of fine, delicate, soft, elastic threads or fibres, and is called **fibrin**.

You see, by stirring, or, as it is frequently called, whipping the blood with the bundle of twigs, you have taken the fibrin out of the blood, and so prevented its clotting.

If you were to take one of the clotted lumps of blood that were spilt on the ground or a bit of the clot from a pail in which the blood had not been whipped, and wash it long enough, you would find at last that all the colour went away from the lump, and you had nothing left but a small quantity of white stringy substance. This white stringy substance is fibrin—exactly the same thing you got on your bundle of twigs.

If the blood is carefully caught in a pail, and afterwards not disturbed at all, it clots into a solid mass. The whole of the blood seems to have changed into a complete jelly; and if you turn it out of the pail, as you may do, it keeps its shape, and gives you quite a mould of the pail, a great trembling red jelly just the shape of the inside of the pail.

But if you were to leave the blood in the pail for a few hours or for a day, you would find, instead of the large jelly quite filling the pail, a smaller but firmer jelly covered by or floating in a colourless or very pale yellow liquid. This smaller, firmer jelly, which in the course of a day or so would get still firmer and smaller, would in fact go on shrinking in size, you may still call the **clot**; the clear fluid in which it is floating is called **serum**.

What has taken place is as follows. Soon after blood is shed there is formed in it a something which was not present in it before. This something, which we call **fibrin**, starts as a multitude of fine tender threads which run in all directions through the mass of blood, forming a close network everywhere. So the blood is shut up in an immense number of little chambers formed by the meshes of the fibrin; and it is this which makes it seem a jelly. But each thread of fibrin as soon as it is formed begins to shrink, and the blood in each of these little chambers is squeezed by the shrinking of its walls of fibrin, and tries to make its way out. The corpuscles get caught in the meshes, but all the rest of the blood passes between the threads and comes out on the top and sides of the pail. And this goes on until you have left in the clot very little besides corpuscles entangled in a network of fibrin, and all the rest of the blood has been squeezed outside the clot, and is then called serum. **Serum, then, is blood out of which the corpuscles have been strained by the process of clotting.**

Now I dare say you are ready to ask the question, If blood clots so readily when it is shed, why does it not clot inside the body? Why is our blood ever fluid? This is rather a difficult question to answer. When blood is shed from the warm body it soon gets cool. But it does not clot and become solid because it gets cool, as ordinary jelly does. If you keep it from getting cool it clots all the same, in fact quicker, and if kept cold enough will not clot at all. Nor does it clot when shed, because it has become still, and is no longer rushing round through the blood-vessels. Nor is it because it is exposed to the air. Perhaps we don't know exactly why it is, and you will have much to learn hereafter about the coagulation of blood. All I will say at present is that as long as the blood is in the body there is something at work to keep it from clotting. It does clot sometimes in the body, and blood-vessels get plugged with the clots; but that constitutes a very dangerous disease.

24. Well, blood is thicker than water because it contains solid corpuscles and fibrin. But even the serum, *i.e.* blood out of which both fibrin and corpuscles have been taken, is thicker than water.

You know that if you were to take a basinful of pure water and boil it, it would boil away to nothing. It would all go off in steam. But if you were to try to boil a basinful of serum, you would find several curious things happen.

In the first place you would not be able to boil it at all. Before you got it as hot as boiling water, your serum, which before seemed quite as liquid as water, only feeling a little sticky if you put your finger in it, would all become quite solid. You know the difference between a raw and a boiled egg. The white of the raw egg, though very sticky and ropy, or viscid as it is called, is still liquid; you will find it hard work if you try to cut it with a knife. The white of the hard boiled egg, on the other hand, is quite solid, and you can cut it into ever so thin slices. It has been "set" by boiling. Well, the serum of blood is in this respect very like white of egg. In fact they both contain the same substance, called **albumin**, which has this property of "setting" or becoming solid when heated nearly to boiling-point. Both the serum of blood and white of egg even when "set" are wet, *i.e.* contain a great deal of water. You may dry them in the proper manner into a transparent horny substance. When quite dry they will readily burn. They are therefore things which can be oxidized. When burnt they give off carbonic acid, water, and ammonia; the latter you might easily recognize by its effect on your nose if you were to burn a piece of dried blood in a flame. Now, when I say that **albumin** in burning gives off carbonic acid, water, and ammonia, you know from your Chemistry that it must contain carbon to form the carbonic acid, hydrogen to form water, and nitrogen to form ammonia. It need not contain oxygen, for as you know it could get all the oxygen it wanted from the air; still it does contain some oxygen. **Albumin, then, is an oxidizable or combustible body made up of nitrogen, carbon, hydrogen, and oxygen.** It is important you should remember this; but I will not bother you with how much of each—it is a very complex substance, built up in a wonderful way, far more complex than any of the things you had to learn about in your Chemistry Primer. And this albumin, dissolved in a great deal of water, forms the serum of blood.

I did not say anything about what fibrin was made of; but it, like albumin, is made up of nitrogen, carbon, hydrogen, and oxygen. It is not quite the same thing as albumin, but first cousin to it. There is another first cousin to both of them, also containing nitrogen, carbon, hydrogen, and oxygen, which together with a great deal of water forms muscle; another forms a great part of the red corpuscles; and scattered all over the body in various places, there are first cousins to albumin, all containing nitrogen, carbon, hydrogen, and oxygen, all combustible, and all when burnt giving off carbonic acid, water, and ammonia. All these first cousins go under one name; they are all called **proteids**.

25. Well, then, blood is thicker than water by reason of the proteids in the corpuscles, in the fibrin, and in the serum, but there is something else besides. I will not trouble you with the crowd of things of which there are perhaps just a few grains in a gallon of blood, like the little pinches of things a cook puts into a savoury dish; though, as you go on in your studies, you will find that these, like many other little things in the world, are of great importance.

But I will ask you to remember this. If you take some dried blood and burn it, though you may burn all the proteids (and some other of the trifles I spoke of just now) away, you will not be able to burn the whole blood away. Burn as long as you like, you will always have left a quantity of what you have learnt from your Chemistry to call **ash, and if you were to examine this ash you would find it contained ever so many elements; sulphur, phosphorus, chlorine, potassium, sodium, calcium, and iron, being the most abundant and most important**.

Blood, then, is a very wonderful fluid: wonderful for being made up of coloured corpuscles and colourless fluid, wonderful for its fibrin and power of clotting, wonderful for the many substances, for the proteids, for the ashes or minerals, for the rest of the things which are locked up in the corpuscles and in the serum.

But you will not wonder at it when you come to see that the blood is the great circulating market of the body, in which all the things that are wanted by all parts, by the muscles, by the brain, by the skin, by the lungs, liver, and kidney, are bought and sold. What the muscle wants, it, as we have seen, buys from the blood; what it has done with it sells back to the blood; and so with every other organ and part. As long as life lasts this buying and selling is for ever going on, and this is why the blood is for ever on the move, sweeping restlessly from place to place, bringing to each part the things it wants, and carrying away those with which it has done. When the blood ceases to move, the market is blocked, the buying and selling cease, and all the organs die, starved for the lack of the things which they want, choked by the abundance of things for which they have no longer any need.

We have now to learn how the blood is thus kept continually on the move.

26. You have already learnt to recognize the blood-vessels of the rabbit, and to distinguish two kinds of blood-vessels—the arteries, which in a dead animal generally contain little or no blood, and have rather firm stout walls; and the veins, which are generally full of blood, and have thinner and flabby walls. The arteries when you cut them generally gape and remain open; the veins fall together and collapse. The larger the arteries, the stouter and firmer they are, and the greater the difference between them and the veins.

You have also studied the capillaries in the frog's foot; you have seen that they are minute channels, with the thinnest and tenderest walls, forming a close network in which the smallest arteries end, and from which the smallest veins begin.

You have moreover been told that all over your own body, in every part, there are, though you cannot see them, networks of capillaries like those in the frog's foot which you can see; that all the arteries of your body end in capillaries, and all the veins begin in capillaries. Let me repeat that, one or two structures excepted, there is no part of your body in which, could you put it under a microscope, you would not see a small artery branching out and losing itself in a network of capillaries, out of which, as out of so many roots, a small vein gathers itself together again.

In some places the network is very close, the capillaries lying closer together than even in the frog's foot; in others the network is more open, and the capillaries wider apart; but everywhere, with a few exceptions which you will learn by and by, there are capillaries, arteries, and veins.

Suppose you were a little lone red corpuscle, all by yourself in the quite empty blood-vessels of a dead body, squeezed in the narrow pathway of a capillary, say of the biceps muscle of the arm, able to walk about, and anxious to explore the country in which you found yourself. There would be two ways in which you might go. Let us first imagine that you set out in the way which we will call backwards. Squeezing your way along the narrow passage of the capillary in which you had hardly room to move, you would at every few steps pass, on your right hand and on your left, the openings into other capillary channels as small as the one in which you were. Passing by these you would presently find the passage widening, you would have more room to move, and the more openings you passed, the wider and higher would grow the tunnel in which you were groping your way. The walls of the tunnel would grow thicker at every step, and their thickness and stoutness would tell you that you were already in an artery, but the inside would be delightfully smooth. As you went on you would keep passing the openings into similar tunnels, but the further you went on, the fewer they would be. Sometimes the tunnels into which these openings led would be smaller, sometimes bigger, sometimes of the same size as the one in which you were. Sometimes one would be so much bigger, that it would seem absurd to say that it opened into your tunnel. On the contrary, it would appear to you that you were passing out of a narrow side passage into a great wide thoroughfare. I dare say you would notice that every time one passage opened into another the way suddenly grew wider, and then kept about the same size until it joined the next. Travelling onwards in this way, you would after a while find yourself in a great wide tunnel, so big that you, poor little corpuscle, would seem quite lost in it. Had you anyone to ask, they would tell you it was the main artery of the arm. Toiling onwards through this, and passing a few but for the most part large openings, you would suddenly tumble into a space so vast that at first you would hardly be able to realize that it was the tunnel of an artery like those in which you had been journeying. This you would learn to be the **aorta**, the great artery of all; and a little further on you would be in the heart.

Suppose now you retraced your steps, suppose you returned from the aorta to the main artery of the arm, and thus back through narrower and narrower tunnels till you came again to the spot from which you started, and then tried the other end of the capillary. You would find that that led you also, in a very similar way, into wider and wider passages. Only you could not help noticing that though the inside of all the passages was as smooth as before, the walls were not nearly so thick and stout. You would learn from this that you were in the veins, and not in the arteries. You would meet too with something, the like of which you did not see in the arteries (except perhaps just close to the heart). Every now and then you would come upon what for all the world looked like one of those watch-pockets that sometimes are hung at the head of a bedstead, a watch-pocket with its opening turned the way you were going. This you would find was called a **valve**, and was made of thin but strong membrane or skin. Sometimes in the smaller veins you would meet with one watch-pocket by itself, sometimes with two or even three abreast, and I dare say you would notice that very frequently, directly you had passed one of these valves, you came to a spot where one vein joined another.

Well, but for these differences, your journey along the veins would be very like your journey along the arteries, and at last you would find yourself in a great vein, whose name you would learn to be the **vena cava**, or hollow vein (and because, though there is but one aorta, there are two great "hollow veins," **the superior vena cava** or **upper hollow vein**), and from thence your next step would be into the heart again. So you see, starting from the capillary (you started from a capillary in the arm, but you might have started from any capillary anywhere), whether you go along the arteries or whether you go along the veins, you at last come to the heart.

Before we go on any further we must learn something about the heart.

27. Go and ask the butcher for a sheep's pluck. There will most probably be one hanging up in his shop. Look at it before he takes it down. The hook on which it is hanging has been thrust through the windpipe. You will see that the sheep's windpipe is, like the rabbit's, all banded with rings of cartilage, only very much larger and coarser. Below the windpipe come the spongy lungs, and between them lies the heart, which perhaps is covered up with a skin and so not easily seen. Hanging to the heart and lungs is the great mass of the liver. When you have got the pluck home, cut away the liver, cut away the skin (pericardium, it is called) which is covering the heart, if it has not been cut away already, and lay the lungs out on a table with the heart between them. You will then have something very much like what

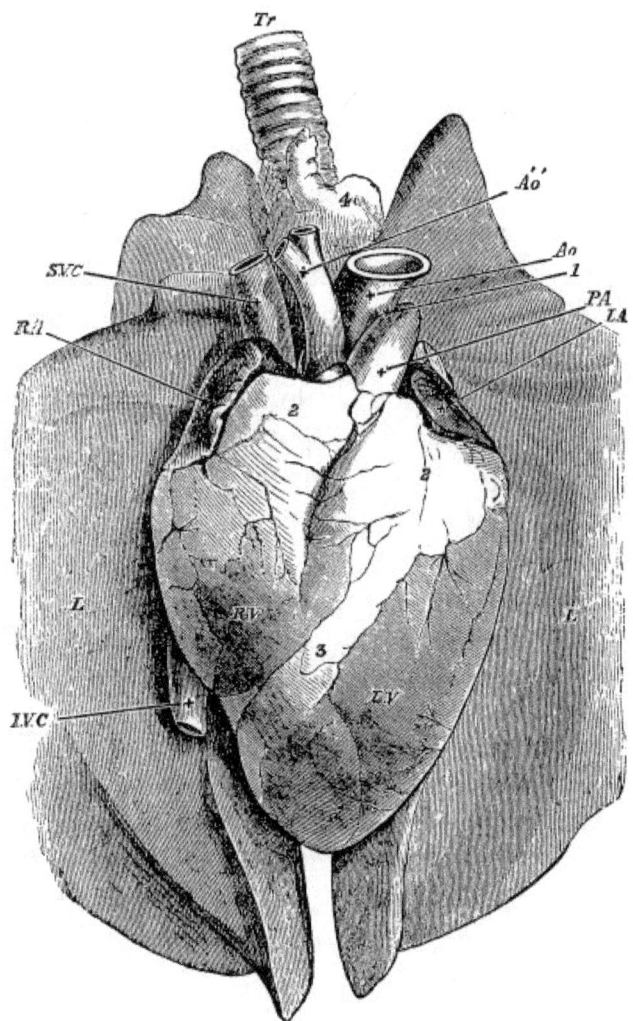

Fig. 5.—Heart of Sheep, as seen after Removal from the Body, lying upon the Two Lungs. The Pericardium has been cut away, but no other Dissection made.

R.A. Auricular appendage of right auricle; L.A. auricular appendage of left auricle; R.V. right ventricle; L.V. left ventricle; S.V.C. superior vena cava; I.V.C. inferior vena cava; P.A. pulmonary artery; Ao, aorta; Áó, innominate branch from aorta dividing into subclavian and carotid arteries; L. lung; Tr. trachea. 1, solid cord often present, the remnant of a once open communication between the pulmonary artery and aorta. 2, masses of fat at the bases of the ventricle hiding from view the greater part of the auricles. 3, line of fat marking the division between the two ventricles. 4, mass of fat covering the trachea.

is represented in Fig. 5. If you could look through the front of your own chest, and see your own heart and lungs in place, you would see something not so very very different.

If now you handle the heart—and if you want to learn physiology you must handle things—you will have no great difficulty in finding the great yellowish tubes marked *Ao* and *Áó* in the figure. Your butcher perhaps may not have cut them across exactly where mine has done, but that will not prevent your recognizing them. You will notice what thick stout walls they have, and how they gape where they are cut. *Ao* is the **aorta**, and *Áó* is a great branch of the aorta, going to the head and neck of one side, perhaps the branch along which we imagined just now that you, a poor little red blood-corpuscle, were travelling. If you were to put a wire through *Áó* you would be able to bring it out through *Ao*, or *vice versâ*. But what is *P.A.* which looks so much like the aorta, though you will find that it has no connection with it? You cannot pass a wire from the aorta into it. It also is an artery, the **pulmonary[2] artery**. We shall have more to say about it directly.

Now try and find what are marked in the figure as *S.V.C.* and *I.V.C.* You will perhaps have a little difficulty in this; and when you have found them you will understand why. They are the great veins of the body. *S.V.C.* is the **superior vena cava**, to form which all the veins from the head and neck and arms join, the vein in which you were journeying a little while ago. *I.V.C.* is the **inferior vena cava**, made out of all the veins from the trunk and the legs. Being veins, they have thin flabby walls; and their sides fall flat together, so that they seem nothing more than little folds of skin, and it becomes very hard to find the passage inside them. But when you have found the opening into them, you will see that you can stretch them out into quite wide tubes, and that their walls, though very much thinner than those of the aorta, so thin indeed that they are almost transparent, are still after a fashion strong. If you put a penholder or thin rod through either you will find that they both seem to lead right into the middle of the heart. With a little care you can pass a rod up *I.V.C.* and bring the end of it out at the top of *S.V.C.* Of course you will understand that both of these veins have been cut off short.

28. Before we go on any further with the sheep's heart, let me tell you something about it, by help of the diagram in Fig. 6, which is meant to represent the whole circulation. You must remember that this figure is a **diagram**, and not a picture; it does not represent the way the blood-vessels are really arranged in your own body. If you had no arms and no legs, and if you only had a few capillaries at the top of your head and at the bottom of your body, it might be more like than it is.

In the centre of the figure is the heart. This you will see is completely divided by an upright partition into two halves, a right half and a left half. Each half is further marked off, but not completely divided, into

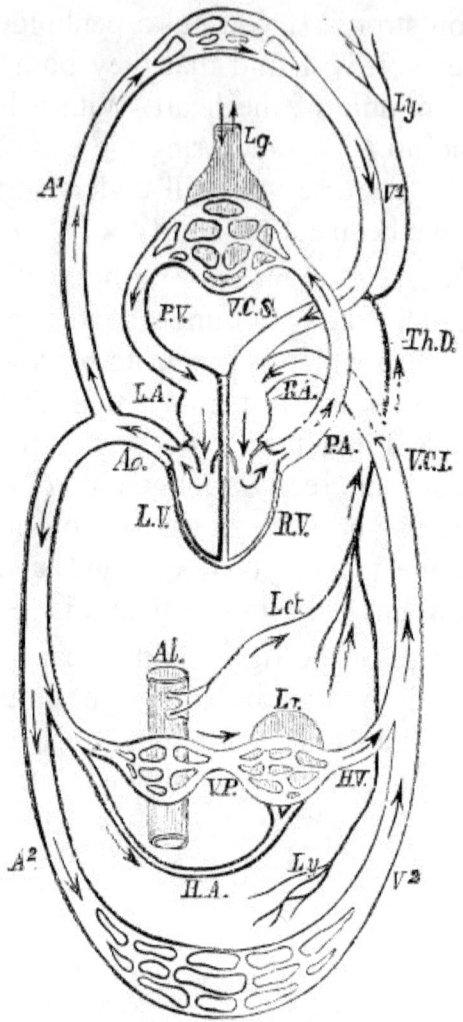

Fig. 6.—Diagram of the Heart and Vessels, with the Course of the Circulation, viewed from behind so that the proper left of the Observer corresponds with the left side of the Heart in the Diagram.

L.A. left auricle; L.V. left ventricle; Ao. aorta; A1. arteries to the upper part of the body; A2. arteries to the lower part of the body; H.A. hepatic artery, which supplies the liver with part of its blood; V2. veins of the upper part of the body; V2. veins of the lower part of the body; V.P. vena portæ; H.V. hepatic vein; V.C.I. inferior vena cava; V.C.S. superior vena cava; R.A. right auricle; R.V. right ventricle; P.A. pulmonary artery; Lg. lung; P.V. pulmonary vein; Lct. lacteals; Ly. lymphatics; Th.D. thoracic duct; Al. alimentary canal; Lr. liver. The arrows indicate the course of the blood, lymph, and chyle. The vessels which contain arterial blood have dark contours, while those which carry venous blood have light contours.

two chambers, an upper chamber and a lower chamber; so that altogether we have four chambers,—two upper chambers, one on each side, marked *R.A.* and *L.A.*, these are called the **right and left auricles**; and two lower chambers, one on each side, marked *R.V.* and *L.V.*, these are called the **right and left ventricles**. The right auricle, *R.A.*, opens in the direction of the arrow into the right ventricle, *R.V.*, the opening being guarded, as we shall see, by a valve. The left auricle, *L.A.*, opens into the left ventricle, *L.V.*, the opening being likewise guarded by a valve; but you have to go quite a roundabout way to get from either the right auricle or ventricle to the left auricle or ventricle. Let us see how we can get round the figure. Suppose we begin with the two tubes marked *V.C.S.* and *V.C.I.*, the walls of which are drawn with thin lines. These both open into the right auricle. They are the vena cava superior and inferior, which you have just made out in the sheep's heart. From the right auricle you pass easily into the right ventricle; thence, following the arrow, the way is straight into the tube marked *P.A.* This is the pulmonary artery, the outside of which you saw in the sheep's heart (Fig. 5, *P.A.*) Travelling along this pulmonary artery, you come to the lungs, and after passing through branches not represented in the figure, picking your way through arteries which continually get smaller and smaller, you find yourself at last in the capillaries of the lungs. Squeezing your way through these, you come out into veins, and gradually advancing through larger and larger veins, you, still following the arrow, find yourself in one of four large veins (only one of them is represented in the diagram) which land you in the left auricle. From the left auricle it is but a jump into the left ventricle. From the left ventricle the way is open, as indicated by the arrow, into the tube marked *Ao.* This is intended to represent the aorta, which you have already seen in the sheep's heart (Fig. 5, *Ao*). It is here drawn for simplicity's sake as dividing into two branches, but you have already been told, and must bear in mind, that it does not in reality divide in this way, but gives off a good many branches of various sizes. However, taking the figure as it stands, suppose we travel along A_2. Following the arrow, and shooting through arteries which continually get smaller and smaller, we come at last to capillaries somewhere, in the skin or in some muscle, or in a bone, or in the brain, or almost anywhere, in fact, in the upper part of the body. Out of the capillaries we pass into veins, which, joining together and so forming larger and larger trunks, bring us at last to the point from which we started, the superior vena cava, *V.C.S.* If we had taken the other road, A_2, we should have passed through capillaries somewhere in the lower part of the body instead of the upper, and come back by the vena cava inferior, *V.C.I.*, instead of the vena cava superior. **Starting from the right auricle, whichever way we took we should always come back to the right auricle again, and in our journey should always pass through the following things in the following order: right auricle, right ventricle, pulmonary artery, arteries, capillaries, and veins of the lungs, pulmonary vein, left auricle, left ventricle, aorta, arteries, capillaries, and veins somewhere in the body, and either superior or inferior vena cava.** That is the course of the circulation. But there is something still to be added. Among the many large branches, not drawn in the diagram, given off by the aorta to the lower part of the body, there are two branches which are drawn and which deserve special notice.

One is a large branch carrying blood to the tube *A.L.*, which is meant in the diagram to stand for the stomach, intestines, and some other organs. This branch, like all other branches of the aorta, divides into small arteries, and these into capillaries, which again are gathered up into veins, forming at last a large vein marked in the diagram *V.P.* and called the **vena portæ** or **portal vein**.

Now the remarkable thing is that this vein does not, like all the other veins, go straight to join the vena cava, but makes for the liver, where it divides into smaller and smaller veins, until at last it breaks completely up in the liver into a set of capillaries again. These capillaries gather once more into veins, forming at last the large trunk, called the **hepatic[3] vein**, *H.V.*, which does what the portal vein ought to have done but did not; it opens straight into the vena cava.

The other branch of the aorta of which we are speaking goes straight to the liver, and is called the **hepatic artery**, *H.A.*: there it breaks up in the liver into small arteries, and then into capillaries, which mingle with the capillaries of the portal vein, and form one system, out of which the hepatic veins spring. So you see it makes a great difference to a red corpuscle which is travelling along the lower part of the aorta A_2, whether it takes a turn into the branch going to the alimentary canal, or whether it goes straight on into, for instance, a branch going to some part of the leg. In the latter case, having got through a set of capillaries, it is soon back into the vena cava and on its road to the heart. But if it takes the turn to the alimentary canal, it finds after it has passed through the capillaries and got into the portal vein, that it has still to go through another set of capillaries in the liver before it can pass through the hepatic vein into the vena cava.

This then is the course of the circulation. Right side of the heart, pulmonary artery, capillaries of the lungs, pulmonary vein, left side of the heart, aorta, capillaries somewhere, sometimes two sets, sometimes one, vena cava, right side of the heart again. A little corpuscle cannot get from the right to the left side of the heart without going through the capillaries of the lungs. It cannot get from the left side of the heart to the right without going through some capillaries somewhere in the body, and if it should happen to take the turn to the stomach, it has to go through two sets of capillaries instead of one.

You see, you really have two circulations, and you have two hearts joined together into one. If you were very skilful you might split the heart in half and pull the two sides asunder, and then you would have one heart receiving all the veins from the body and sending its arteries (branches of the pulmonary artery) all to the lungs, and another heart receiving all the veins from the lungs and sending its arteries (branches of the aorta) all over the body. And you would have two circulations, one through the lungs, and another through the rest of the body, both joining each other. Very often two circulations are spoken of, and because the lungs are so much smaller than the rest of the body, the circulation through the lungs is called the lesser circulation, that through the rest of the body the greater circulation.

29. I have described the circulation as if the blood always went in one direction from the right side of the heart to the left, from arteries to veins, the way the arrows point in the diagram. And so it does. It cannot go the other way round. **Why does it go that way? Why cannot it go the other way round?**

The reasons are to be found partly in the heart, partly in the veins.

In the veins the blood will only pass from the capillaries to the heart. Why not from the heart to the capillaries? You remember the little watch-pocket-like valves, here and there, sometimes singly, sometimes two or three abreast. **You remember that the mouths of the watch-pockets were always turned towards the heart.** Now suppose a crowd of little corpuscles hurrying along a vein towards the heart. When they came to one of these watch-pocket valves they would simply trample it down flat, and so pass over it without hardly knowing it was there, and go on their way as if nothing had happened. But suppose they were journeying the other way, from the heart to the capillaries. When they came to the open mouth of a watch-pocket valve, some of them would be sure to run into the pocket, and then the pocket would bulge out, and the more it bulged out the more blood would run into it, until at last it would be so full of blood that it would press close against the top of the vein, as is shown in Fig. 7 (or, if there were two or three, they would all meet together), and so quite block the vein up. If you doubt this, make a watch-pocket out of a piece of silk or cotton, fasten it on to a piece of brown paper, and roll the paper up into a tube, so that the valve is nicely inside the tube. If you pour some peas down the tube with the mouth of the valve looking away from you, they will run through at once; but if you try to pour them the other way, your tube will soon be choked, and if you carefully unroll the tube you will find the watch-pocket crammed full of peas.

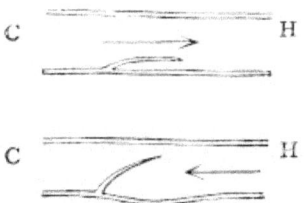

Fig. 7.—Diagrammatic Sections of Veins with Valves.

In the upper, the blood is supposed to be flowing in the direction of the arrow, towards the heart; in the lower, the reverse way. C, capillary side; H, heart side.

The valves in the veins, then, let the blood pass easily from the capillaries to the heart, but won't let it go the other way. If you bare your arm you may see some of the veins in the skin, in which the blood is running up from the hand towards the shoulder. If with your finger you press one of these veins back towards the hand it will swell up, and if you look carefully you may see little knots here and there caused by the bulging out of the watch-pocket valves. If you press it the other way, towards the elbow, you will empty it easily, and if with another finger you prevent the blood getting into it from behind, that is from the hand, the vein will remain empty a very long time.

The presence of valves in the veins, then, is one reason why the blood moves in one direction, but other reasons, and these the chief ones, are to be found in the heart.

Let us now go back to the sheep's heart.

30. You know from the diagram that the two great veins, the superior and inferior vena cava, open into the right auricle. If you slit up these two veins in the sheep's heart, you will find that they end by separate openings in a small cavity, the inside of which is for the most part smooth, and the walls of which, made, as you will at once see, of muscle, are not very thick. This small cavity is the right auricle, shown in Fig. 8, *R.A.*, where the great veins have not been slit up, but the front of the auricle has been cut away. In this auricle, beside the openings into the two great veins and another one which belongs to a vein coming from the heart itself (Fig. 8, *b*) there is quite a large one, leading straight downwards, into which you

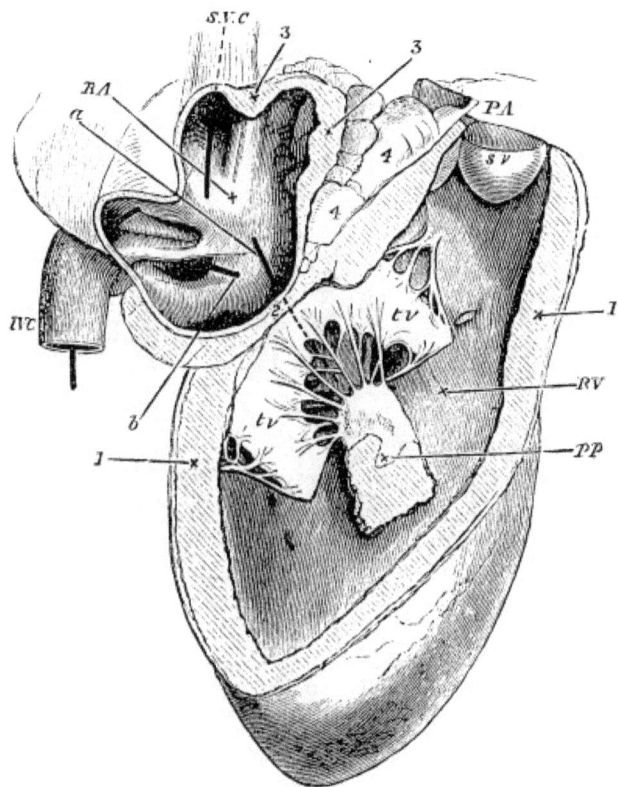

Fig. 8.—Right Side of the Heart of a Sheep.

R.A. cavity of right auricle; S.V.C. superior vena cava; I.V.C. inferior vena cava; (a piece of whalebone has been passed through each of these;) a, a piece of whalebone passed from the auricle to the ventricle through the auriculo-ventricular orifice; b, a piece of whalebone passed into the coronary vein.

R.V. cavity of right ventricle; tv, tv, two flaps of the tricuspid valve: the third is dimly seen behind them, the a, piece of whalebone, passing between the three. Between the two flaps, and attached to them by chordæ tendineæ, is seen a papillary muscle, PP, cut away from its attachment to that portion of the wall of the ventricle which has been removed. Above, the ventricle terminates somewhat like a funnel in the pulmonary artery, P.A. One of the pockets of the semilunar valve, sv, is seen in its entirety, another partially.

1, the wall of the ventricle cut across; 2, the position of the auriculo-ventricular ring; 3, the wall of the auricle; 4, masses of fat lodged between the auricle and pulmonary artery.

can put your three fingers. This is the opening into the right ventricle; and you will have no difficulty in putting your fingers from the auricle into the ventricle and bringing them out again.

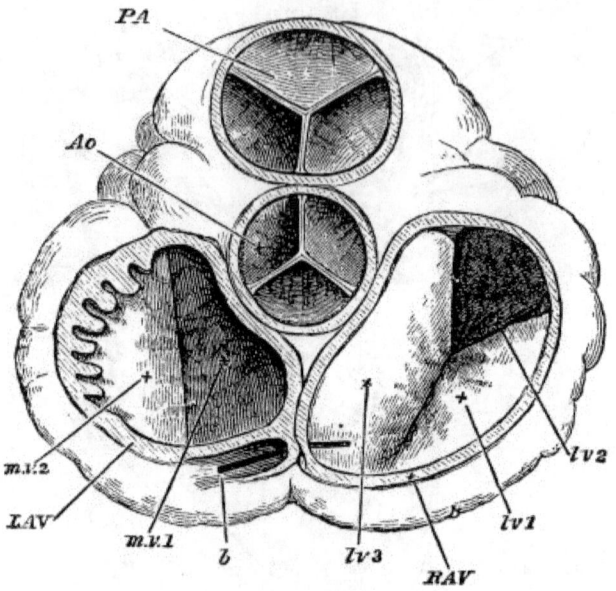

Fig. 9.—The Orifices of the Heart seen from above, the Auricles and Great Vessels being cut away.

P.A. pulmonary artery, with its semilunar valves; Ao. aorta, do.

R.A.V. right auriculo-ventricular orifice with the three flaps (lv. 1, 2, 3) of tricuspid valve.

L.A.V. left auriculo-ventricular orifice, with m.v. 1 and 2, flaps of mitral valve; b, piece of whalebone passed into coronary vein. On the left part of L.A.V. the section of the auricle is carried through the auricular appendage; hence the toothed appearance due to the portions in relief cut across.

But hold the heart in one hand with the auricle upwards, and try to pour some water into the ventricle. The first few spoonfuls will go in all right, and then you will see some thin white skin or membrane come floating up into the opening and quite block up the entrance from the auricle into the ventricle; the

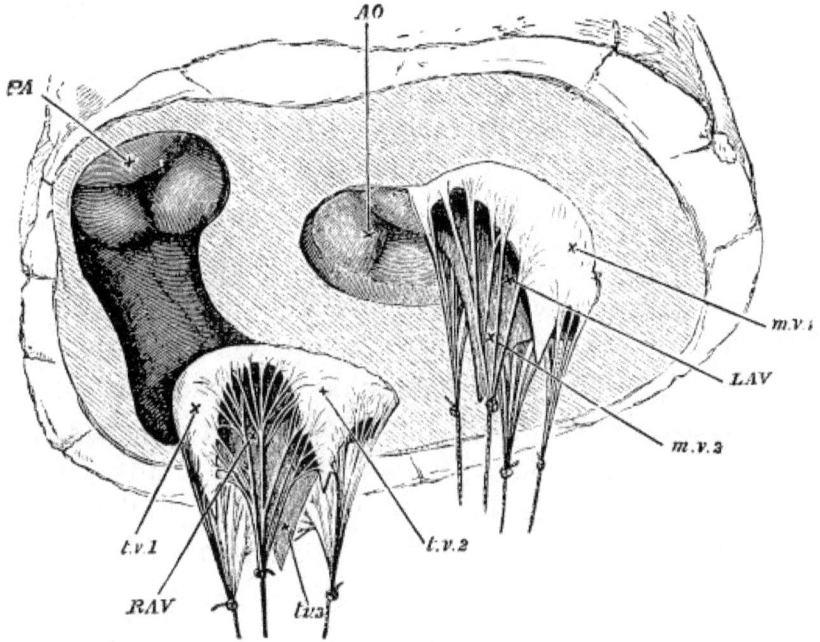

Fig. 10.—View of the Orifices of the Heart from below, the whole of the Ventricles having been cut away.

R.A.V. right auriculo-ventricular orifice surrounded by the three flaps, t.v. 1, t.v. 2, t.v. 3, of the tricuspid valve; these are stretched by weights attached to the chordæ tendineæ.

L.A.V. left auriculo-ventricular orifice surrounded in same way by the two flaps, m.v. 1, m.v. 2, of mitral valve; P.A. the orifice of pulmonary artery, the semilunar valves having met and closed together; Ao. the orifice of the aorta with its semilunar valves. The shaded portion, leading from R.A.V. to P.A., represents the funnel seen in Fig. 8.

water will immediately fill the auricle and run over. If you look at the membrane carefully as it comes bulging up, you will notice that it is made up of three pieces joined together as is shown in Fig. 9 (*lv.* 1, *lv.* 2, *lv.* 3). These three pieces form the valve between the right auricle and ventricle, called the **tricuspid**, or three-peaked valve. Why it is so called you will understand if you lay open the right ventricle by cutting with a pair of scissors from the auricle into the ventricle along the side of the heart, or by cutting away the front of the ventricle as has been done in Fig. 8. You will then see that the valve is made up of three little triangular flaps, which grow together round the opening with their points hanging down into the cavity of the ventricle (Fig. 10, *t.v.*) They do not, however, hang quite loosely. You will notice fastened to the sides of the flaps, thin delicate threads, the other ends of which are fastened to the sides of the ventricle, and often to little fleshy projections called papillary muscles (Fig. 8, *P.P.*)

How do these valves act? In this way. When the ventricle is empty, and blood or water or any other fluid is poured into it from the auricle, the valves are pushed on one side against the walls of the ventricle, and thus there is a great wide opening from the auricle into the ventricle. But as the ventricle fills, the blood or water gets behind the flaps and floats them up towards the auricle. The more fluid in the ventricle the higher they float, until when the ventricle is quite full they all meet together in the middle of the opening between the auricle and ventricle and completely block it up. But why do they not turn right over into the auricle, and so open up again the wrong way? Because of those little threads (the *chordæ tendineæ*, as they are called) which fasten them to the walls of the ventricle. The flaps float back until these threads are stretched quite tight, and the threads are just long enough to let the flaps reach to the middle of the opening, but no further. The tighter the threads are stretched the closer the flaps fit together, and the more completely do they block the way from the ventricle back into the auricle.

The tricuspid valve, then, lets blood flow easily from the right auricle into the right ventricle, but prevents it flowing from the ventricle into the auricle.

31. Now look at the cavity of the ventricle. Its walls are fleshy, that is muscular, and you will notice that they are much stouter and thicker than those of the auricle. Besides the opening from the auricle there is but one other, which is at the top of the ventricle, side by side

with the former. If you put a penholder or your finger through this second opening, you will find that it leads into the large vessel which you have already learnt to recognize as the pulmonary artery (Fig. 5, *P.A.*)

Slit up the pulmonary artery from the ventricle with a pair of scissors, as has been done in Fig. 8, *P.A.* You will notice at once the line where the red soft flesh of the muscular ventricle leaves off, and the yellow firmer material of which the artery is made begins. Just at that line you will see a row of three (perhaps you may have cut one of the three with your scissors) most beautiful, watch-pocket valves, made on just the same principle as those in the veins, only larger, and more exquisitely finished. These are called **semilunar valves**, because each pocket is of the shape of a half-moon. Lift them up carefully and see how tender and yet how strong they are. There is no need to tell you the use of these. You know it at once. **They are to let the blood flow from the ventricle into the pulmonary artery, and to prevent the blood going back from the artery into the ventricle.**

On the right side of the heart we have, then, two great valves, the tricuspid valve between the auricle and the ventricle, and the semilunar valve between the ventricle and the pulmonary artery. These let the blood flow easily one way, but not the other. If you doubt this, try it. Put a tube into either the superior or inferior vena cava of a fresh heart, tying the other vena cava and another tube into the pulmonary artery. If with a funnel you pour water into the tube in the vein, it will run through auricle and ventricle and out through the tube of the pulmonary artery as easily as possible; but if you try to pour water the other way down the pulmonary artery, you will find you cannot do it; the tube gets blocked directly, and only a few drops come back through the heart into the vein.

Now slit up the pulmonary artery as far as you can, and note when you cut it how stout and firm are its walls. You will find that it soon divides into two branches, one for the right lung, one for the left. Each of these, when it gets to the lung, divides into branches, and these again into others, as far as you can follow them. You know from what you have learnt already that these branches end in capillaries all over the lungs.

32. Not far from the two main branches of the pulmonary artery you will find, covered up perhaps with fat and other matters, some tubes which you will at once recognize as veins, and if you open any one of these you will find that you can put a thin rod into it, and that it leads in one direction to the lungs, and in the other into the left side of the heart. These are the *pulmonary veins*, and if you slit them right up you will find they open (by four openings) into a cavity on the left side of the heart, almost exactly like that cavity on the right side which we called the right auricle (Fig. 11). This cavity is, in fact, the **left auricle**; out of it there is an opening into the left ventricle, very like the opening from the right auricle into the right ventricle. It too is guarded by flap valves, exactly like the tricuspid valve, only there are but two flaps instead of three (Fig. 9, *m.v.* 1, *m.v.* 2). Hence this valve is called the **bicuspid**, or more frequently the **mitral** valve. Its flaps have little threads by which they are fastened to the walls of the ventricle, and in fact, except for there being two flaps instead of three, the mitral valve is exactly like the tricuspid valve, and acts exactly the same way.

If you cut with a pair of scissors from the auricle into the ventricle, you will find the left ventricle (Fig. 11) very much like the right ventricle, only its walls are very much thicker, so much thicker that the left ventricle takes up the greater part of the heart. You will see this if you now look at the outside of a fresh heart.

The auricles are so small and so covered up by fat that from the outside you can hardly see them at all. What you chiefly see are two little fleshy corners, one of each auricle (Fig. 5, *R.A. L.A.*), often called "the auricular appendages." By far the greater part is taken up by the ventricles—and if you look you will see a band of fat slanting across the heart (Fig. 5, 3). This marks the line of the fleshy division, or **septum** as it is called, between the two ventricles. You will notice that the point or apex of the heart belongs altogether to the left ventricle.

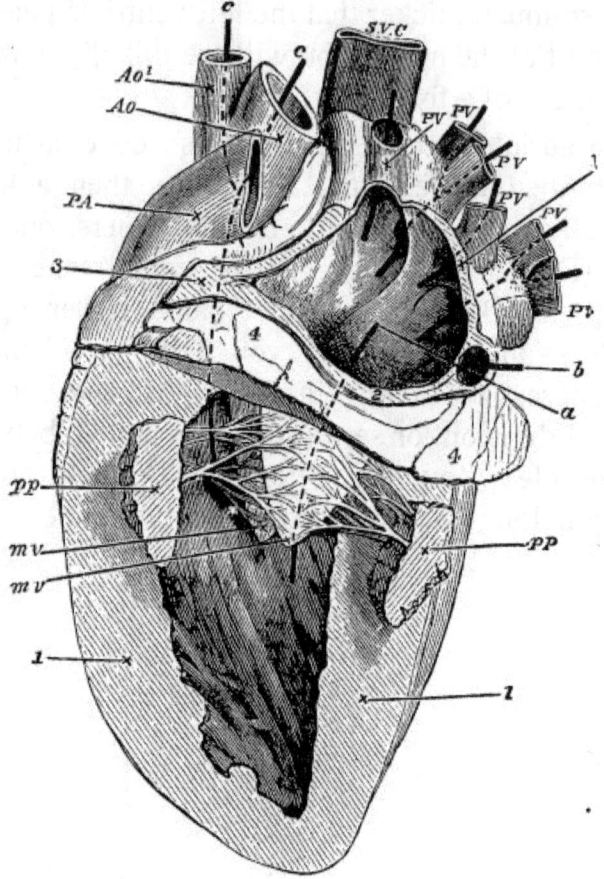

Fig. 11.—Left Side of the Heart of a Sheep (laid open).

P.V. pulmonary veins opening into the left auricle by four openings, as shown by the styles or pieces of whalebone placed in them: a, a style passed from auricle into ventricle through the auriculo-ventricular orifice; b, a style passed into the coronary vein, which, though it has no connection with the left auricle, is, from its position, necessarily cut across in thus laying open the auricle.

M.V. the two flaps of the mitral valve (drawn somewhat diagrammatically): pp, papillary muscles, belonging as before to the part of the ventricle cut away; c, a style passed from ventricle in Ao. aorta; Ao2. branch of aorta (see Fig. 5, Aó); P.A. pulmonary artery; S.V.C. superior vena cava.

1, wall of ventricle cut across; 2, wall of auricle cut away around auriculo-ventricular orifice; 3, other portions of auricular wall cut across; 4, mass of fat around base of ventricle (see Fig. 5, 2).

To return to the inside of the left ventricle. Up at the top of the ventricle, close to the opening from the auricle, there is one other opening, and only one. If you put your finger into this, you will find that it leads into a tube which first of all dips under or behind the pulmonary artery and then comes up and to the front again. This tube is what you already know as the **aorta.** If you slit it up from the ventricle (and to do this you must cut through the pulmonary artery), you will find that on the left side, as on the right, the red fleshy wall of the ventricle suddenly changes into the yellow firm wall of the artery, and that just at this line there are three semilunar valves exactly like those in the pulmonary artery.

On the left side of the heart, then, we have also two valves, the mitral between the auricle and the ventricle, and the semilunar between the ventricle and the aorta. These let the blood pass one way and not the other. You can easily drive fluid from the pulmonary veins through auricle and ventricle into the aorta, but you cannot send it back the other way from the aorta.

These then are the reasons why the blood will only pass one way, the way I said it did. There are sets of valves opening one way and shutting the other. These valves are the tricuspid between the right auricle and right ventricle, the pulmonary semilunar valves between the right ventricle and the pulmonary artery, the mitral valve between the left auricle and the left ventricle, the aortic semilunar valves between the left ventricle and the aorta, and the valves which are scattered among the veins of the body. Of these by far the most important are the valves in the heart: they do the chief work; those in the veins do little more than help.

33. Well, then, we understand now, do we not? why the blood, if it moves at all, moves in the one way only. There still remains the question, **Why does the blood move at all?**

You know that during life it does keep moving. You have seen it moving in the web of a frog's foot—and whenever any part of the body can be brought under the microscope, the same rush of red corpuscles through narrow channels may be seen. You know it moves because when you cut a blood-vessel the blood runs out. If you cut an artery across, the blood gushes out from the end which is nearest the heart; if you cut a vein across, the blood comes most from the end nearest the capillaries. If you want to stop an artery bleeding, you tie it between the cut and the heart; if you want to stop a vein bleeding, you tie it between the cut and the capillaries. You understand now why there is this difference between a cut artery and a cut vein. And you see that this is by itself a proof that the blood moves in the arteries from the heart to the capillaries, and in the veins from the capillaries to the heart.

The blood is not only always moving, but moves very fast. It flies along the great arteries at perhaps ten inches in a second. Through the little bit of capillaries along which it has to pass it creeps slowly, but manages sometimes to go all the way round from vein to vein again in about half a minute.

It is always moving at this rapid rate, and when it ceases to move, you die.

What makes it move?

Suppose you had a long thin muscle, fastened at one end to something firm, and with a weight hanging at the other end. You know that every time the muscle contracted it would pull on the weight and draw it up. But suppose, instead of hanging a weight on to the muscle, you wrapped the muscle round a bladder full of water. What would happen then each time the muscle contracted? Why, evidently it would squeeze the bladder, and if there were a hole in the bladder some of the water would be squeezed out. That is just what takes place in the heart. You have already learnt that the heart is muscular. Each cavity of the heart, each auricle, and each ventricle is, so to speak, a thin bag with a number of muscles wrapped round it. In an ordinary muscle of the body, the bundles of fibres of which the muscle is made up are placed carefully and regularly side by side. You can see this very well in a round of boiled beef, which is little more than a mass of great muscles running in different directions. You know that if you try to cut a thin slice right across the round, at one part your carving-knife will go "with the grain" of the meat, *i.e.* you will cut the fibres lengthways; at another part it will go "against the grain," *i.e.* you will cut the fibres crossways. In both parts, the bundles of fibres will run very regularly. But in the heart the bundles are interlaced with each other in a very wonderful fashion, so that it is very difficult to make out the grain. They are so arranged in order that the muscular fibres may squeeze all parts of each bag at the same time.

Each cavity of the heart, then, auricle or ventricle, is a thin bag with a network of muscles wrapped round it, and each time the muscles contract they squeeze the bag and try to drive out whatever is in it. There are more muscles in the ventricles than in the auricles, and more in the left ventricle than in the right, for we have already seen how much thicker the ventricles are than the auricles, and the left ventricle than the right; and the thickness is all muscle.

And now comes the wonderful fact. These muscles of the auricles and ventricles are always at work contracting and relaxing, shortening and lengthening, of their own accord, as long as the heart is alive. The biceps in your arm contracts only when you make it contract. If you keep quiet, your arm keeps quiet and your biceps keeps quiet. But your heart never keeps quiet. Whether you are awake or whether you are asleep, whether you are running about or lying down quite still, whatever you are doing or not doing, as long as you are alive your heart keeps on steadily at work. Every second, or rather oftener, there comes a short sharp squeeze from the auricles, from both exactly at the same time, and just as the auricles have finished their squeeze, there comes a great hug from the ventricles, from both at the same time, but a much stronger hug from the left than from the right; and then for a brief space there is perfect quiet. But before the second has quite passed away, the auricles have begun again, and after them the ventricles once more, and thus the contracting and relaxing of the walls of the heart's cavities, this beat of the heart as it is called, this short snap of the two auricles, this longer, steadier pull of the two ventricles, have gone on in your own body since before you were born, and will go on until the moment comes when friends gathering round your bedside will say that you are "gone."

34. But how does this beat of the heart make the blood move? Let us see.

Remember that you have, or when you are grown up will have, bottled up in the closed blood-vessels of your body about 12 lbs. of blood. You have seen that the heart and the blood-vessels form a system of closed tubes; the walls are in some places, in the capillaries for instance, very thin, but they are sound and whole—and though the road is quite open from the capillaries through the veins, heart, and arteries to the capillaries again, there is no way out of the tubes except by making a breach somewhere in the walls.

This closed system of heart and tubes is pretty well filled by the 12 lbs. of blood.

What then must happen each time the heart contracts?

Let us begin with the right ventricle. Suppose it is full of blood. It contracts. The blood in it, squeezed on all sides, tries to go back into the right auricle, but the tricuspid flaps have been driven back and block the way. The more the blood presses on them, the tighter they become, and the more completely they shut out all possibility of getting into the auricle.

The way into the pulmonary artery is open, the blood can go there. But stay, the artery is already full of blood, and so are the capillaries and veins in the lung. Yes, but the artery will stretch ever so much. Take a piece of pulmonary artery, and having tied one end, pump or pour water into the other; you will see how much it will stretch. Into the pulmonary artery, then, goes the blood, stretching it in order to find room. As the ventricle squeezes and squeezes, until its walls meet in the middle, all the blood that was in it finds its way out into the artery. But the beat of the ventricle soon ceases, the squeeze is over and gone, and back tumbles the blood into the ventricle, or would tumble, only the first few drops that shoot backwards are caught by the watch-pocket semilunar valves. Back fly these valves with a sharp click (for the things of which we are speaking happen in a fraction of a second), and all further return is cut off. The blood has been squeezed out of the ventricle, and is safely lodged in the pulmonary artery.

But the pulmonary artery is ever so much on the stretch. It was fairly full before it received this fresh lot of blood; now it is over-full—at least that part of it which is nearest to the heart is over-full. What happens next? What happens when you stretch a piece of india-rubber and then let it go? It returns to its former size. The ventricle has stretched the piece of pulmonary artery near it, beyond the natural size, and then (when it ceased to contract) has let it go. Accordingly the piece of pulmonary artery tries to return to its former size, and since it cannot send the blood back to the ventricle, squeezes it on to the next piece of the artery nearer the capillaries, stretching that in turn.

This again in turn sends it on the next piece—and so on right to the capillaries. The over-full pulmonary artery, stretched to hold more than it fairly can, empties itself through the capillaries into the pulmonary veins until it is not more than comfortably full. But the pulmonary veins also are already full,—what are they to do? To empty the surplus into the left auricle. Oftener than every second there will come a time when they can do so.

For at the same time that the right ventricle pumped a quantity of blood into the pulmonary artery and safely lodged it there, the left ventricle pumped a like quantity into the aorta, safely lodged it there, and was left empty itself. But just at that moment the left auricle began to contract and to squeeze the blood that was in it.

Where could that blood go? It could not go back into the pulmonary veins, for they were already full, and the blood in them was being pressed behind by the over-full pulmonary arteries. But it could pass easily into the empty ventricle—and in it tumbled, the mitral flaps readily flying back and opening up a wide way. And so the auricle emptied itself into the ventricle. But now the auricle ceases to contract—its walls no longer squeeze— it is empty and wants filling, and so comes the moment when the pulmonary veins can pour into it the blood which has been driven into them by the over-full pulmonary artery.

Thus the right ventricle drives the blood into the over-full pulmonary artery, the pulmonary artery overflows into the pulmonary veins, the pulmonary veins carry the surplus to the empty left auricle, the left auricle presses it into the empty left ventricle, the left ventricle pumps it into the aorta—(the stretching of the aorta and of its branches is what we call the pulse)—the over-full aorta overflows just as did the pulmonary artery, through the capillaries of the body into the great venæ cavæ— through these the blood falls into the empty right auricle, the right auricle drives it into the empty right ventricle, and the full right ventricle is the point at which we began.

Thus the alternate contractions of auricles and ventricles, thanks to the valves in the heart and in the veins, pump the blood, stroke by stroke, through the wide system of tubes; and thus in every capillary all over the body we find blood pressed upon behind by over-full arteries, with a way open to it in front, thanks to the auricles, which are, once a second or oftener, empty and ready to take up a fresh supply from the veins. Thus it comes to pass that every little fragment of your body is bathed by blood, which a few moments ago was in your heart, and a few moments before that was in some other part of your frame. Thus it is that no part of your body can keep itself to itself; the blood makes all things common as it flies from spot to spot. The red corpuscle that a minute ago was in your brain, is now perhaps in your liver, and in another minute may be in a muscle of your arm or in a bone of your leg: wherever it goes it has something to bring, and something to fetch. A restless heart is for ever driving a busy blood, which wherever it goes buys and sells, making perhaps an occasional bargain as it shoots along the great arteries and great veins, but busiest of all as it lingers in the narrow pathways of the capillaries.

35. When you look down upon a great city from a high place, as upon London from St. Paul's, you see stretching below you a network of streets, the meshes of which are filled with blocks of houses. You can watch the crowds of men and carts jostling through the streets, but the work within the houses is hidden from your view. Yet you know that, busy as seems the street, the turmoil and press which you see there are but tokens of the real business which is being carried on in the house.

So is it with any piece of the body upon which you look through the microscope. You can watch the red blood jostling through the network of capillary streets. But each mesh bounded by red lines is filled with living flesh, is a block of tiny houses, built of muscle, or of skin, or of brain, as the case may be. You cannot *see* much going on there, however strong your microscope; yet that is where the chief work goes on. In the city the raw material is carried through the street to the factory, and the manufactured article may be brought out again into the street, but the din of the labour is within the factory gates. In the body the blood within the capillary is a stream of raw material about to be made muscle, or bone, or brain, and of stuff which, having been muscle, or bone, or brain, is no longer of any use, and is on its way to be cast out. The actual making of muscle, or of bone, or of brain, is carried on, and the work of each is done, outside the blood, in the little plots of tissue into which no red corpuscle comes.

The capillaries are closed tubes; they keep the red corpuscles in their place. But their walls are so thin and delicate that they let the watery plasma of the blood, the colourless fluid in which the corpuscles float, soak through them into the parts inside the mesh. You probably know that many things will pass through thin skins and membranes in which no holes can be found even after the most careful search. If you put peas into a bladder and tie the neck, the peas will not get out until the bladder is untied or torn. But if you were to put a solution of sugar or of salt into the bladder, and place the bladder with its neck tied ever so tightly in a basin of pure water, you would find that very soon the water in the basin would begin to taste of sugar or salt—and that without your being able to discover any hole, however small, in the bladder. By putting various substances in the bladder, you will find that solid particles and things which will not dissolve in water keep inside the bladder, whereas sugar and salt, and many other things which dissolve in water, will make their way through the bladder into the water outside, and will keep on passing until the water in the basin is as strong of sugar or salt as the water in the bladder. This property which membranes such as a bladder have of letting certain substances pass through them is called **osmosis**. You will at once see how important a part it plays in your own body. It is by osmosis chiefly that the raw nourishing material in the blood gets into the little islets of flesh lying, as we have seen, in the meshwork of the capillaries. It is by osmosis chiefly that the worn-out stuff from the same islets gets back into the blood. It is by osmosis chiefly that food gets out of the stomach into the blood. It is by osmosis chiefly that the waste, worn-out matters are drained away from the blood, and so cast out of the body altogether. By osmosis the blood nourishes and purifies the flesh. By osmosis the blood is itself nourished and kept pure.

There are two chief things by which the blood, and through the blood the body, is nourished. These are food and air. The air we have always with us, we have no need to buy it or toil for it; hence we take it as we want it, a little at a time, and often. We gather up no store of it; and cannot bear the lack of it for more than a few moments.

For our food we have to labour; we store it up in our bodies from time to time, at intervals of hours, in what we call meals, and can go hours or even days without a fresh supply.

Let us first of all see how the blood, and, through the blood, the body, is nourished by air.

HOW THE BLOOD IS CHANGED BY AIR: BREATHING.
§ VI.

36. I have already said, perhaps more than once, that our muscles burn, burn in a wet way without giving light. And when I say our muscles, I might say our whole body, some parts burning more fiercely than others.

You have learnt from your Chemistry Primer (Art. 2, p. 2) what happens when a candle is placed in a closed jar of pure air. The oxygen gets less, carbonic acid comes in its place, and after a while the candle goes out for want of oxygen to carry on that oxidation which is the essence of burning. You also know that exactly the same thing would happen if you were (only you need not do it) to put a bird or a mouse in the jar instead of a candle. The oxygen would go, carbonic acid would come, and the little flame of life in the mouse would flicker and go out, and after a while its body would be cold.

But suppose you were to put a fish or a snail in a jar of pure fresh water, and cork the jar tight. There seems at first sight to be no air in the jar. But there is. If you were to take that fresh water, and put it under an air-pump, you could pump bubbles of air out of it; and if you were to examine these bubbles you would find them to contain oxygen and nitrogen, with very little carbonic acid. **The water contains dissolved air.** After the fish or the snail had been some time in the jar, you would see its flame of life flicker and die out, just like that of the bird in air; and if you then pumped the air out of the water you would find that the oxygen was nearly gone and that carbonic acid had come in its place.

You see, then, that air can be breathed, as we call it, even when it is dissolved in water.

Now to return to our muscle. When you were watching the circulation in the frog's foot, you could tell the artery from the vein, because in the artery the blood was flowing *to* the capillaries, and in the vein *from* them. Both artery and vein were rather red, and of about the same tint of colour. But if you could see in your own body a large artery going to your biceps muscle, and a large vein coming away from it, you would be struck at once with the difference of colour between them. The artery would look bright scarlet, the vein a dark purple; and if you were to prick both, the blood would gush from the artery in a bright scarlet jet, and bubble from the vein in a dark purple stream. And wherever you found an artery and a vein (with a great exception of which I shall have to speak directly), the blood in the artery would be bright scarlet, and that in the vein dark purple. Hence we call the bright scarlet blood which is found in the arteries **arterial blood,** and the dark purple blood which is found in the veins **venous blood.**

What is the difference between the two? If you were to pump away at some arterial blood, as you did at the water in which you put your fish, you would be able to obtain from it some air, or, more correctly, some gas; a great deal more gas, in fact, than you did from the water. A pint of blood would yield you half a pint of gas. This gas you would find on examination not to be air, *i.e.* not made up of a great deal of nitrogen and the rest oxygen. (Chemical Primer, Art. 9.) There would be very little nitrogen, but a good deal of oxygen, and still more carbonic acid.

If you were to pump away at some venous blood you would get about as much gas, but it would be very different in composition. The little nitrogen would remain about the same, but the oxygen would be about half gone, while the carbonic acid would be much increased.

This, then, is one great difference (for there are others) between venous and arterial blood, that while **both contain, dissolved in them, oxygen, nitrogen, and carbonic acid, venous blood contains less oxygen and more carbonic acid than arterial blood.**

37. In passing through the capillaries on its way to the vein, the blood in the artery has lost oxygen and gained carbonic acid. Where has the oxygen gone to? Whence comes the carbonic acid? To and from the islets of flesh between the capillaries, to the bloodless muscular fibre or bit of nerve or skin which the blood-holding capillaries wrap round. The oxygen has passed from the blood within the capillaries to the flesh outside; from the flesh outside the carbonic acid has passed to the blood within the capillaries. And this goes on all over the body. Everywhere the flesh is breathing blood, is breathing gas dissolved in the blood, just as a fish breathes water, *i.e.* breathes the air dissolved in the water.

Goes on everywhere with one great exception. There is one great artery, with its branches, in which blood is not bright, scarlet, arterial, but dark, purple, venous. There are certain great veins in which the blood is not dark, purple, venous, but bright, scarlet, arterial. You know which they are. The pulmonary artery and the pulmonary veins. The blood in the pulmonary veins contains more oxygen and less carbonic acid than the blood in the pulmonary artery. **It has lost carbonic acid and gained oxygen, as it passed through the capillaries of the lungs.**

38. What are the lungs? As you saw them in the rabbit, or as you may see them in the sheep, they are shrunk and collapsed. We shall presently learn why. But if you blow into them through the windpipe, which divides into branches, one for either lung, you can blow them out ever so much bigger. They are in reality bladders which can be filled with air, but which, left to themselves, at once empty themselves again.

They are bladders of a peculiar construction. Imagine a thick short bush or tree crowded with leaves; imagine the trunk and the branches, small and great, down to the veriest twigs, all hollow; imagine further that the leaves themselves were little hollow bladders, stuck on to the smallest hollow twigs, and made of some delicate, but strong and exceedingly elastic, substance. If you blew down the trunk you might stretch and swell out all the hollow leaves; when you left off blowing they would all fall together, and shrink up again.

Around such a framework of hollow branches called **bronchial-tubes**, and hollow elastic bladders called **air-cells**, is wrapped the intricate network of pulmonary arteries, veins, and capillaries, in such a way that each air-cell, each little bladder, is covered by the finest and most close-set network of capillaries, very much as a child's india-rubber ball is covered round with a network of string. Very thin are the walls of the air-cell, so thin that the blood in the capillary is separated from the air in the air-cell by the thinnest possible sheet of finest membrane. As the dark purple blood rushes through the crowded network, its carbonic acid escapes through this thin membrane, from the blood into the air, and oxygen slips from the air into the blood.

Thus the dark purple venous blood coming along the pulmonary artery, as it glides in the pulmonary capillaries along the outside of the inflated air-cells, by loss of carbonic acid and gain of oxygen is changed into the bright scarlet blood of the pulmonary veins.

This then is the mystery of our constant need of air. **The flesh of the body of whatever kind, everywhere all over the body, breathes blood, making pure arterial blood venous and impure, all over the body except in the lungs, where the blood itself breathes air, and changes from impure and venous to pure and arterial.**

39. Through the capillaries of the muscle a stream of blood is ever flowing so long as life lasts and the heart has power to beat; every instant a fresh supply of bright, pure, arterial blood comes to take the place of that which has become dark, venous, and impure. Without this constant renewal of its blood the muscle would be choked, and its vital flame would flicker and die out.

In the lungs, the air filling the air-cells would if left to itself soon lose all its oxygen and become loaded with carbonic acid; and the blood in the capillaries of the lungs would no longer be changed from venous to arterial, but would travel on to the pulmonary vein as dark and impure as in the pulmonary artery. **Just as the blood in the muscle must be constantly renewed, so must the air in the lungs be continually changed.**

How is this renewal of the air in the lungs brought about?

In the dead rabbit you saw the lungs, shrunk, collapsed, emptied of much of their air, and lying almost hidden at the back of the chest (Fig. 1, *G.G.*) The cavity of the chest seemed to be a great empty space, hardly half filled by the lungs and heart. But this is quite an unnatural condition of the lungs. Take another rabbit, and before you touch the chest at all, open the abdomen and remove all its contents—stomach, liver, intestines, &c. You will then get a capital view of the **diaphragm**, which as you already know forms a complete partition between the chest and the belly. You will notice that it is arched up towards the chest, so that the under surface at which you are looking is quite hollow. If you hold the rabbit up by its hind legs with its head hanging down, and pour some water into the abdomen, quite a little pool will gather in the shallow cup of the diaphragm.

In the rabbit the diaphragm is very transparent; you can see right through it into the chest, and you will have no difficulty in recognizing the pink lungs shining through it. **You will notice that they cover almost all the diaphragm—in fact they fill up the whole of the cavity of the chest that is not occupied by the heart.**

If you seize the diaphragm carefully in the middle with a pair of forceps, and pull it down towards the abdomen, you will find that you cannot create a space between the lungs and the diaphragm, but that the lungs follow the diaphragm, and are quite as close to it when it is pulled down as when it is drawn up.

In other words, when the diaphragm is arched up as you find it on opening the abdomen, the lungs quite fill the chest; and when the diaphragm is drawn down and the cavity of the chest made bigger, the lungs swell out so that they still fill up the chest.

40. How do they swell out? By drawing air in through the windpipe. If you listen, you will perhaps hear the air rush in as you pull the diaphragm down—and if you tie the windpipe, or quite close up the nose and mouth, you will find it much harder to pull down the diaphragm, because no fresh air can get into the lungs.

Now prick a hole through the diaphragm into the cavity of the chest, without wounding the lungs. You will hear a sudden rush of air, and the lungs will shrink up almost out of sight. They are no longer close against the diaphragm as they were before; and if you open the chest you will find that they have shrunk to the back of the thorax as you saw them in the first rabbit. The rush of air is partly a rush of air out of the lungs, and partly a rush of air into the chest between the chest walls and the outside of the lungs.

But before you lay open the chest, pull the diaphragm up and down as you did before you made the hole in the diaphragm. You will find that you have no effect whatever on the lungs. They remain perfectly quiet, and do not swell up at all. By working the diaphragm up and down, you only drive air through the hole you have made, **in and out of the cavity of the chest**, not **in and out of the lungs** as you did before.

We see then that the **chest is an air-tight chamber**, and that the lungs, when the chest walls are whole, **are always on the stretch**, are on the stretch even when the diaphragm is arched up as high as it can go.

Why is it that the lungs are thus always on the stretch? Because the chest is air-tight, so that no air can get in between the outside of the lungs and the inside of the chest wall. You know from your Physics Primer (Art. 29, p. 34) that the atmosphere is always pressing on everything. It is pressing on all parts of the rabbit; it presses on the inside of the windpipe and on the inside of the lungs. It presses on the outside of the abdomen, and so presses on the under surface of the diaphragm, and drives it up into the chest as far as it will go. But it will not go very far, because its edges are fastened to the firm walls of the chest. The air also presses on the outside of the chest, but cannot squeeze that much, because its walls are stout.

If the walls of the chest were soft and flabby, the atmosphere would squeeze them right up, and so through them press on the outside of the lungs; since they are firm it cannot. The chest walls keep the pressure of the atmosphere off the outside of the lungs.

The lungs then are pressed by the atmosphere on their insides and not on their outsides; and it is this inside pressure which keeps them on the stretch or expanded. When you blow into a bladder, you put it on the stretch and expand it because the pressure of your breath inside the bladder is greater than the pressure of the atmosphere outside the bladder. If, instead of making the pressure inside *greater* than that outside, you were to make the pressure outside *less* than that inside, as by putting the bladder under an air-pump, you would get just the same effect; you would expand the bladder. That is just what the chest walls do; they keep the pressure outside the lungs less than that inside the lungs, and that is why the lungs, as long as the chest walls are sound, are always expanded and on the stretch.

When you make a hole into the chest, and let the air in between the outside of the lungs and the chest wall, the pressure of the atmosphere gets at the outside of the lungs; there is then the same atmospheric pressure outside as inside the lungs; there is nothing to keep them on the stretch, and so they shrink up to their natural size, just as does the bladder when you leave off blowing into it, or when you take it out of the air-pump.

When before you made the hole in the diaphragm you pulled the diaphragm down, you **still further lessened the pressure on the outside of the lungs**; hence the pressure inside the lungs caused them to swell up and follow the diaphragm. But this put the lungs still more on the stretch, so that when you let go the diaphragm and ceased to pull on it, the lungs went back again to their former size, emptying themselves of part of their air and pulling the diaphragm up with them. When there is a hole in the chest wall, pulling the diaphragm down does not make any difference to the pressure outside the lungs. They are then always pressed upon by the same atmospheric pressure inside and outside, and so remain perfectly quiet.

When in an air-tight chest the diaphragm is pulled down, the pressure of the atmosphere drives air into the lungs through the windpipe and swells them up. When the diaphragm is let go, the stretched lungs return to their former size, emptying themselves of the extra quantity of air which they had received.

Suppose now the diaphragm were pulled down and let go again regularly every few seconds: what would happen? Why, every time the diaphragm went down a certain quantity of air would enter into the lungs, and every time it was let go that quantity of air would come out of the lungs again.

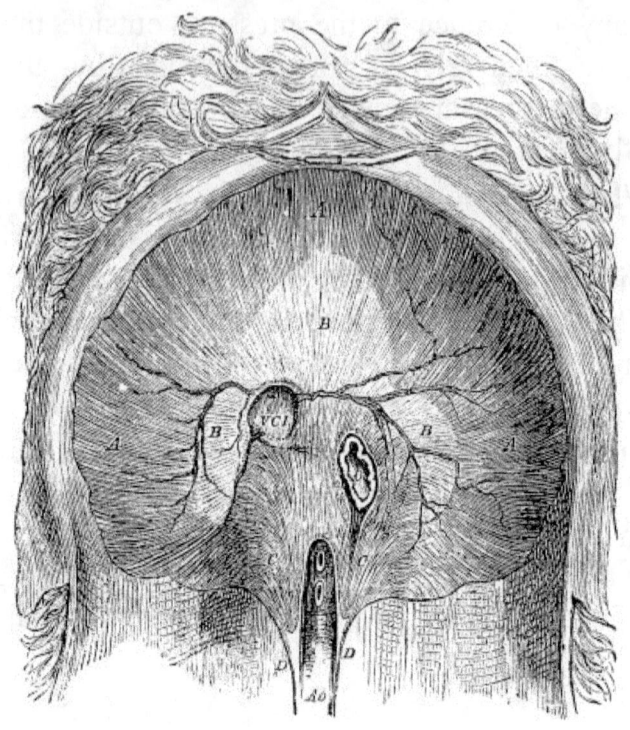

Fig. 12.—The Diaphragm of a Dog viewed from the Lower or Abdominal Side.

V.C.I. the vena cava inferior; O. the œsophagus; Ao. the aorta; the broad white tendinous middle (B) is easily distinguished from the radiating muscular fibres (A) which pass down to the ribs and into the pillars (C D) in front of the vertebræ.

This is what does take place in breathing or respiration. Every few seconds, about seventeen times a minute, the diaphragm does descend, and a quantity of air rushes into the lungs through the windpipe. This is called **inspiration**. As soon as that has taken place, the diaphragm ceases to pull downwards, the stretched lungs return to their former size, carrying the diaphragm up with them, and squeeze out the extra quantity of air. This is called **expiration**.

As the diaphragm descends it presses down on the abdomen; when it ceases to descend, the contents of the abdomen help to press it up. If you place your hand on your stomach, you can feel the abdomen bulging out each time the diaphragm descends in inspiration, and going in again each time the diaphragm returns to its place in expiration.

41. But what causes the diaphragm to descend?

If you look at the diaphragm of the rabbit (or of any other animal) a little carefully, you will see that it is in reality a flat thin muscle, rather curiously arranged; for the red fleshy muscular fibres are on the outside all round the edge (Fig. 12, *A* and *C*), while the centre *B* is composed of a whitish transparent tendon. These muscular fibres, like all other muscular fibres, have the power of contracting. What must happen when they contract and become shortened?

When these muscular fibres are at rest, as in the dead rabbit, the whole diaphragm is arched up, as we have seen, towards the thorax, somewhat as is shown in Fig. 13, *B*. It is partly pushed up by all the contents of the abdomen (for the cavity of the abdomen, you will remember, is quite filled by the liver, stomach, intestines, and other organs), partly pulled up by the lungs, which, as we know, are always on the stretch. When the muscular fibres contract, they pull at the central tendon (just as the biceps pulls at its lower tendon), **and pull the diaphragm flat**; and some of the fibres, such as those at *C*, Fig. 12, also pull it **down. The diaphragm during its contraction is flattened and descends**, somewhat as is shown in Fig. 13, *A*.

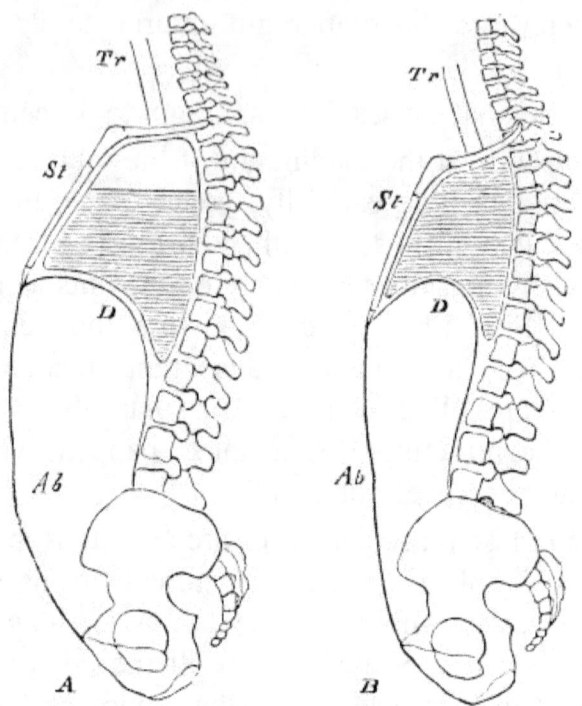

Fig. 13.—Diagrammatic Sections of the Body in

A. inspiration; B. expiration. Tr. trachea; St. sternum; D. diaphragm; Ab. abdominal walls. The shading roughly indicates the stationary air. The unshaded portion at the top of A is the tidal air.

The descent of the diaphragm in inspiration is caused by a contraction of its muscular fibres. During expiration the diaphragm is at rest; its muscular fibres relax; and it goes up because it is partly drawn up by the lungs, partly pushed up by the contents of the abdomen.

42. Other structures besides the diaphragm assist in pumping air in and out of the lungs. By the action of the diaphragm the chest is alternately lengthened and shortened. But if you watch anyone, and especially a woman, breathing, you will notice that with every breath the chest rises and falls; the front of the chest, the sternum, as you have learnt to call it, comes forward and goes back; and a little attention will convince you that it comes forward during inspiration, *i.e.* while the diaphragm is descending, and falls back during expiration. But this coming forward of the sternum means a widening of the chest from back to front, and the falling back of the sternum means a corresponding narrowing. So that while the chest is being lengthened by the descent of the diaphragm, it is also being widened by the coming forward of the sternum. In inspiration the lungs are expanded not only downwards, by the movement of the diaphragm, but also outwards, by the movement of the walls of the chest.

What thrusts forward the sternum? If you were to watch closely the sides of the chest of a very thin person, you would be able to notice that at every breathing in, at every inspiration, the ribs are pulled up a little way. Now, each rib is connected with the backbone behind by a joint, and is firmly fastened to the sternum in front by cartilage (see Frontispiece). If you were to fasten a piece of string to the middle of one of the ribs and to pull it, you would find you were working on a lever, with the fulcrum at the backbone, with the weight acting at the sternum, and the power at the point where your string was tied. Every time you pulled the string the rib would move on its fulcrum at the backbone, in such a way that the front end of the rib would rise up, and the sternum would be thrust out a little. When you left off pulling, the sternum, which in being thrust forward had been put on the stretch, would sink back, and the rib would fall down to its previous position.

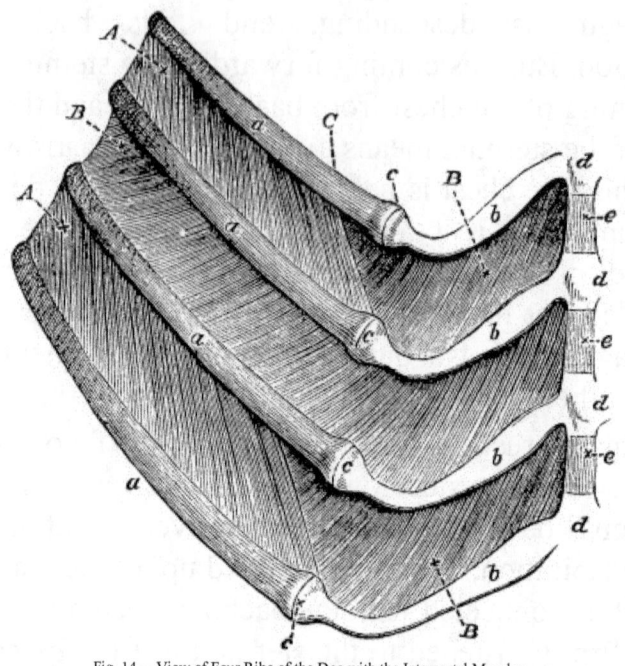

Fig. 14.—View of Four Ribs of the Dog with the Intercostal Muscles.

a. The bony rib; b, the cartilage; c, the junction of bone and cartilage; d, unossified; e, ossified, portions of the sternum. A. External intercostal muscle. B. Internal intercostal muscle. In the middle interspace, the external intercostal has been removed to show the internal intercostal beneath it.

Between the ribs are certain muscles called **intercostal muscles** (Fig. 14). The exact action of these you will learn at some future time. Meanwhile it will be enough to say that they act like the piece of string we are speaking of. **When they contract, they pull up the ribs and thrust out the sternum; when they leave off contracting, the ribs and sternum fall back to their previous position.**

There are many other muscles which help in breathing, especially in hard or deep breathing, but it will be sufficient for you to remember that in ordinary breathing there are two chief movements taking place exactly at the same time, by means of which air is drawn into the chest, both movements being caused by the contraction of muscles. First, the diaphragm contracts and flattens itself, making the chest deeper or longer; secondly, at the same time the ribs are raised and the sternum thrust out by the contraction of the intercostal muscles, making the chest wider. But as the chest becomes wider and longer, the lungs become wider and longer too. In order to fill up the extra room thus made in the lungs, air enters into them through the windpipe. This is **inspiration**. But soon the diaphragm and the intercostal muscles cease to contract; the diaphragm returns to its arched condition, the ribs sink down, the sternum falls back, and the extra air rushes back again out of the lungs through the windpipe. This is **expiration**. An inspiration and an expiration make up a whole breath; and thus we breathe some seventeen times in every minute of our lives.

43. But what makes the diaphragm and intercostal muscles contract and rest in so beautifully regular a fashion? The biceps of the arm, we saw, was made to contract by our will. It is not our will, however, which makes us breathe. We breathe often without knowing it; we breathe in our sleep when our will is dead; we breathe whether we will or no, because we cannot help it. We can quicken our breathing, we can take a short or deep breath as we please, we can change our breathing by the force of our will; but the breathing itself goes on without, and in spite of, our will. It is **an involuntary act.**

Though breathing is not an effort of the will, it is an effort of the brain; an effort, too, of one particular part of the brain, that part where the brain joins on to the spinal cord. Nerves run from the diaphragm and the intercostal and other muscles through the spinal cord, to this part of the brain. And seventeen times a minute a message comes down along these nerves, from the brain, bidding them contract; they obey, and you breathe. Why and how that message comes, you will learn at some future time. When your head is cut off, or when that part of the brain which joins on to the spinal cord is injured by accident or made powerless by disease, the message ceases to be sent, and you cease to breathe.

44. At every breath, then, a certain quantity of air goes in and out of the chest; but only a small quantity. **You must not think the lungs are quite emptied and quite filled at each breath.** On the contrary, you only take in each time a mere handful of air, which reaches about as far as the large branches of the windpipe, and does not itself go into the air-cells at all. This is often called **tidal** air; and the rest of the air in the lungs, which does not move, is often called the **stationary** air (see Fig. 13).

How then does the carbonic acid at the bottom of the lungs get out? How do the capillaries in the air-cells get their fresh oxygen?

The stationary air mingles with the tidal air at every breath. If you want to ventilate a room, you are not obliged to take a pair of bellows and drive out every bit of the old air in the room, and supply its place with new air: it will be enough if you open a window or a door and let in a draught of pure air across one corner, say, of the room. That current of pure air flowing across the corner will mingle with all the rest of the air until the whole air in the room becomes pure; and the mingling will take place very quickly. So it is in the lungs. The tidal air comes in with each inspiration as pure air from without; but before it comes out at the next expiration it gives up some of its oxygen to the stationary air, and robs the stationary air of some of its carbonic acid. For each breath of tidal air the stationary air is so much the better, having lost some of its carbonic acid and gained some fresh oxygen. The tidal air rapidly purifies the stationary air, and the stationary air purifies the blood.

Thus it comes to pass that the tidal air, which at each pull of the diaphragm and push of the sternum goes into the chest as pure air with twenty-one parts oxygen to seventy-nine parts nitrogen in every hundred parts, comes out, when the diaphragm goes up and the sternum falls back, as impure air with only sixteen parts oxygen, but with five parts carbonic acid to seventy-nine of nitrogen. That lost oxygen is carried through the stationary air to the blood in the capillaries, and the gained carbonic acid came through the stationary air from the blood in the capillaries. So each breath helps to purify the blood, and the pumping of air in and out of the chest changes the impure, hurtful, venous, to pure, refreshing, arterial blood; the blood breathes air in the lungs, that all the body may in turn breathe blood.

HOW THE BLOOD IS CHANGED BY FOOD:
DIGESTION. § VII.

45. The blood is not only purified by air, it is also renewed and made good by food. The food we eat becomes blood. But our food, though frequently moist, is for the most part solid. We cut it into small pieces on the plate, and with our teeth we crush and tear it into still smaller morsels in our mouth. Still, however well chewed, a great deal of it, most of it in fact, is swallowed solid. In order to become blood it must first be dissolved. It is dissolved in the alimentary canal, and we call the dissolving **digestion**. Let us see how digestion is carried on.

Your skin, though sometimes quite moist with perspiration, is as frequently quite dry. The inside of your mouth is always moist—very frequently quite filled with fluid; and even when you speak of it as being dry, it is still very moist. Why is this? The inside of your mouth is also very much redder than your skin. The redness and the moisture go together.

In speaking of the capillaries, I said that almost all parts of the body were completely riddled with them, but **not quite all**. A certain part of the skin, for instance, has no capillaries or blood-vessels at all. You know that where your skin is thick, you can shave off pieces of skin without "fetching blood;" if your

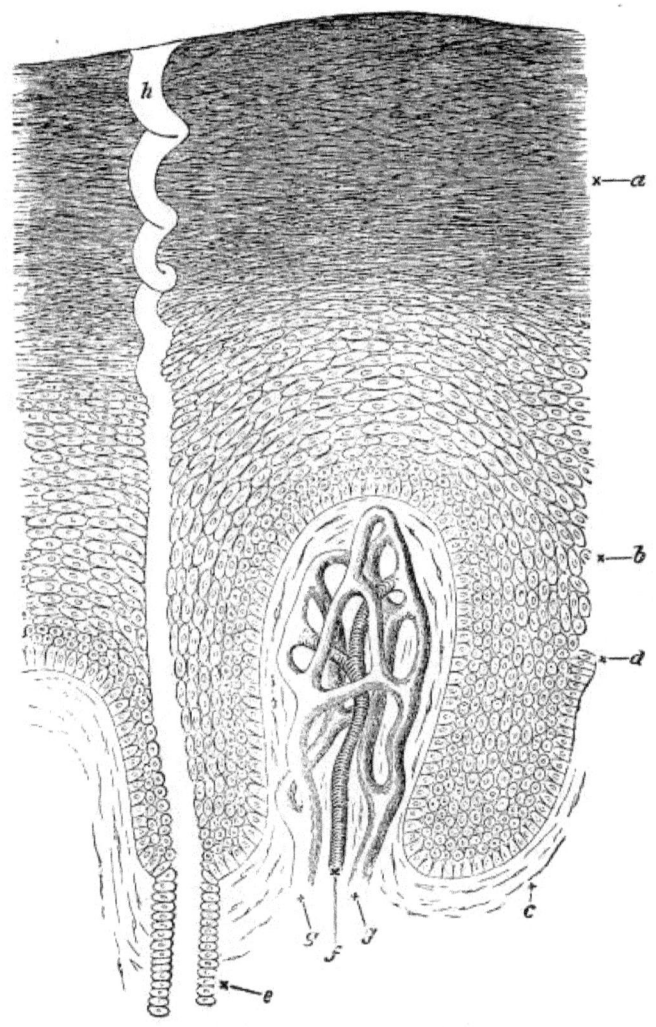

Fig. 15.—Section of Skin, highly magnified.

a, horny epidermis; b, softer layer; c, dermis; d, lowermost vertical layer of epidermic cells; e, cells lining the sweat duct continuous with epidermic cells; h, corkscrew canal of sweat duct. To the right of the sweat duct the dermis is raised into a papilla, in which the small artery, f, breaks up into capillaries, ultimately forming the veins, g.

knife were very sharp and you very skilful, you might do the same in every part of your skin. If you were to put some of the skin you had thus cut off under the microscope, you would find that it was made up of little scales. And if you were to take a very thin upright slice running through the whole thickness of the skin, and examine that under a high power of the microscope, you would find that the skin was made up of two quite different parts or layers, as shown in Fig. 15. The upper layer, *a, b*, is nothing but a mass of little bodies packed closely together. At the top they are pressed flat into scales, but lower down they are round or oval, and at the same time soft. They are called **cells**. As you advance in your study of Physiology you will hear more and more about cells. This layer of cells, either soft and round, or flattened and dried into scales, is called the **epidermis**. No blood-vessel is ever found in the epidermis, and hence, when you cut it, it never bleeds. As long as you live it is always growing. The top scales are always being rubbed off. Whenever you wash your hands, especially with soap, you wash off some of the top scales; and you would soon wash your skin away, were it not that new round cells are always being formed at the bottom of the epidermis, along the line at *d* (Fig. 15), and always moving up to the top, where they become dried into scales. Thus the skin, or more strictly the epidermis, is always being renewed. Sometimes, as after scarlet fever, the new skin grows quickly, and the old skin comes away in great flakes or patches.

The lower layer below the epidermis is what is called the **dermis**, or **true skin**. This is full of capillaries and blood-vessels, and when the knife or razor gets down to this, you bleed. It is not made up of cells like the epidermis, but of that fibrous substance which you early learnt to call connective tissue (see p. 9). Its top is rarely level, but generally raised into little hillocks, called **papillæ**, as in the figure; the epidermis forming a thick cap over each papillæ, and filling up the hollows between them. Most of the papillæ are full of blood-vessels.

Now, then, I think you will understand why your skin is not red, but flesh-coloured, and why it is generally dry. The true skin under the epidermis is always moist, because of the blood-vessels there; the waste and fluid parts of the blood pass readily through the walls of the capillaries, as you have learnt, by osmosis, and so keep everything round them moist. But this moisture is not enough to soak through the thick coating of epidermis, and so the top part of the epidermis remains dry and scaly.

The true skin underneath the epidermis is always red; you know that if you shave off the surface of your skin anywhere, it gets redder and redder the deeper you go down, even though you do not fetch blood. It is red because of the immense number of capillaries, all full of red blood, which are crowded into it. When you look at these capillaries through a great thickness of epidermis, the redness is partly hidden from you, as when you put a sheet of thin white paper over a red cloth, and the skin seems pink or flesh-coloured; and where the epidermis is very thick, as at the heel, the skin is not even pink, but white or yellow, more or less dirty according to circumstances.

46. But if the moist true skin is thus everywhere covered by a thick coat of epidermis, which keeps the moisture in, how is it that the skin is nevertheless sometimes quite moist, as when we perspire?

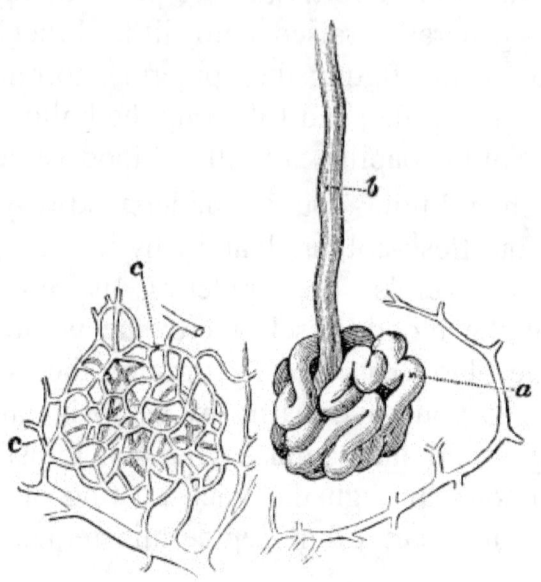

Fig. 16.—Coiled end of a Sweat Gland, Epithelium not shown.
a, the coil; b, the duct; c, network of capillaries, inside which the duct gland lies.

If you look at Fig. 15, you will see that the epidermis is at one point pierced by a canal (*h*) running right through it. You will notice that this canal is not closed at the bottom of the epidermis, but runs right into the dermis or true skin, where the canal becomes a tube, with just one layer (*e*) of cells, like the cells of the epidermis, for its walls. There is no room in Fig. 15 to show what becomes of this tube, but it runs some way down under the skin all among the blood-vessels, and then twisting itself up into a knot, ends blindly, as is shown in Fig. 16, where *b* is a continuation on a smaller scale of the same tube which is seen in Fig. 15. This knot is covered by a close network of capillaries, which at *c* are supposed to be unravelled and taken away from the knotted tube in order to show them. The capillaries, you will understand, though inside the knot, are always outside the tube. If you were to drop a very diminutive marble in at *h* (Fig. 15), it would rattle down the corkscrew passage through the thick epidermis, shoot down the straight tube *b* (Fig. 16), and roll through the knot *a*, until it came to rest at the blind end of the tube. Along its whole course it would touch nothing but cells, like the cells of the epidermis, a single layer of which forms the walls of the tube where it runs below the epidermis. If it got lodged at *h* (Fig. 15), or got lodged in the knot at *a* (Fig. 16), it would in both cases be touching epidermic cells. But there would be this great difference. At *h* it would be ever so far removed from any blood capillary; at *a* it would only have to make its way through a thin layer of single cells, and it would be touching a capillary directly. At *h* it might remain dry for some time; at *a* it would get wet directly, for there is nothing to prevent the fluid parts of the blood oozing out through the thin wall of the capillaries, and so through the thin wall of the tube into the canal of the tube, on to the marble.

In fact, the inside of the knot is always moist and filled with fluid. When the capillaries round the knot get over-full of blood, as they often do, a great deal of colourless watery fluid passes from them into the tube. The tube gets full, the fluid wells up right into the corkscrew portion in the thickness of the epidermis, and at last overflows at the mouth of the tube over the skin. We call this fluid sweat or **perspiration**. We call the tube with its knotted end **a gland**; and we call the act by which the colourless fluid passes out of the blood capillaries into the canal of the tube, **secretion**. We speak of the **sweat gland secreting sweat out of the blood brought by the capillaries which are wrapped round the gland**.

47. Now we can understand why the inside of the mouth is red and moist. The mouth has a skin just like the skin of the hand. There is an outside epidermis, made up of cells and free from capillaries, and beneath that a dermis or true skin crowded with capillaries. Only the epidermis of the mouth is ever so much thinner than that of the hand. The red capillaries easily shine through it, and their moisture can make its way through. Hence the mouth is red and moist. Besides there are many glands in it, something like the sweat gland, but differing in shape; these especially help to keep it moist.

Because it is always red and moist and soft, the skin of the inside of the mouth is generally not called a skin at all, but **mucous membrane**, and the upper layer is not called epidermis, but **epithelium**. You will remember, however, that a mucous membrane is in reality a skin in which the epidermis is thin and soft, and is called epithelium.

The mouth is the beginning of the alimentary canal. Throughout its whole length the alimentary canal is lined by a skin or mucous membrane like that of the mouth, only over the greater part of it the epithelium is still thinner than in the mouth, and indeed is made up of a single layer only of cells. The whole of the inside of the canal is therefore red and moist, and whatever lies in the canal is separated by a very thin partition only from the blood in the capillaries, which are found in immense numbers in the walls of the canal. The alimentary canal is, as you know, a long tube, wide at the stomach but narrow elsewhere. In all parts of its length the tube is made up of mucous membrane on the inside, and on the outside of muscles, differing somewhat from the muscles of the body and of the heart, but having the same power of contracting, and by contracting of squeezing the contents of the tube, just as the muscles of the heart squeeze the blood in its cavities. The muscles, and especially the mucous membrane, are crowded with blood-vessels.

Though the epithelium of the mucous membrane is very thin, the mucous membrane itself is thick, in some places quite as thick as the skin of the body. This thickness is caused by its being **crowded with glands**. In the skin the sweat glands are generally some little distance apart, but in the mucous membrane of the stomach and of the intestines they are packed so close together, that the membrane seems to be wholly made up of glands.

These glands vary in shape in different parts. Nowhere are they exactly like the sweat glands, because none of them are long thin tubes coiled up at the end in a knot, and none of them have a great thickness of epidermis to pass through. Most of them are short, rather wide tubes; some of them are branched at the deep end. They all, however, resemble the sweat glands in being tubes or pouches closed at the bottom but open at top, lined by a single layer of cells, and wrapped round with blood capillaries. From these capillaries, a watery fluid passes into the tubes, and from the tubes into the alimentary canal. This watery fluid is, however, of a different nature from sweat, and is not the same in all parts of the canal. The fluid which is, as we say, secreted by the glands in the walls of the stomach is an **acid fluid**, and is called **gastric juice**; that by the glands in the walls of the intestines is **an alkaline fluid**, and is called **intestinal juice**.

48. But besides these glands in the mucous membrane of the mouth, the stomach, and the intestines, there are other glands, which seem at first sight to have nothing to do with the mucous membrane.

Beneath the skin, underneath each ear, just behind the jaw, is a soft body, which ordinarily you cannot feel, but which, when inflamed by what is called "the mumps," swells up into a great lump. In a sheep's head you would find just the same body, and if you were to examine it you would notice fastened to it a fleshy cord running underneath the skin across the cheek towards the mouth. By cutting the cord across you would discover that what seemed a cord was in reality a narrow tube coming from the soft body we are speaking of and opening into the mouth. Just close to the soft body this tube divides into two smaller tubes, these divide again into still smaller ones, or give off small branches; all these once more divide and branch like the boughs of a tiny tree; and so they go on branching and dividing, getting smaller and smaller, until they end in fine tubes with blind swollen ends. All the tubes, great and small, are lined with epithelium and wrapped round with blood-vessels, and being packed close together with connective tissue, make up the soft body we are speaking of. This body is in fact a **gland**, and is called a **salivary gland**; as you see it is not a simple gland like a sweat gland, but is made up of a host of tube-like glands all joined together, and hence is called a **compound gland**. Being placed far away from the mouth, it has to be connected with the cavity of the mouth by a long tube, which is called its **duct**. You cannot fail to notice how like such a gland is, in its structure, to a lung. The lung is in fact a gland secreting carbonic acid: and the duct of the two lungs is called the trachea. The salivary gland beneath the ear is called the **parotid gland**; there is another very similar one underneath the corner of the jaw on either side, called the **submaxillary gland**. By each of them a watery fluid is secreted, which, flowing along their ducts into the mouth and being there mixed with the moisture secreted by the other glands in the mouth, is called **saliva**.

In the cavity of the abdomen lying just below the stomach is a much larger but altogether similar compound gland called **the pancreas**, which pours its secretion called **pancreatic juice** into the alimentary canal just where the small intestine begins (Fig. 17, *g.*)

That large organ the liver, though the plan of its construction is not quite the same as that of the pancreas or salivary glands, as you will by and by learn, is nevertheless a huge gland, secreting from the blood capillaries into which the portal vein (see p. 62) breaks up, a fluid called **bile** or **gall**, which by a duct, the **gall duct**, is poured into the top of the intestine (Fig. 17, *e*). When bile is not wanted, as when we are fasting, it turns off by a side passage from the duct into the gall-bladder (Fig. 17, *f*), to be stored up there till needed.

49. What are the uses of all these juices and secretions? To dissolve the food we eat.

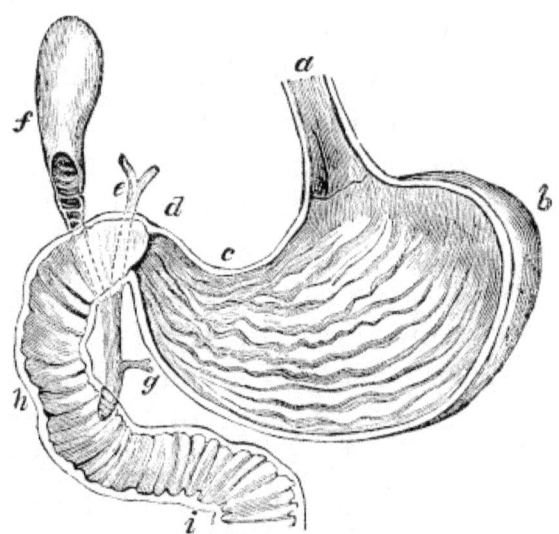

Fig. 17.—The Stomach laid open behind.

a, the œsophagus or gullet; b, one end of the stomach; d, the other end joining the intestine; e, gall duct; f, the gall-bladder; g, the pancreatic duct; h, i, the small intestine.

We eat all manner of dishes, but in all of them that are worth eating we find the same kind of things, which we call **food-stuffs**.

We eat various kinds of meat; but all meats are made up chiefly of two things: the substance of the muscular fibre, which you have already learnt is a **proteid** matter containing nitrogen, and the **fat** which wraps round the lean muscular flesh. Now, proteids are, when cooked, insoluble in water (see p. 49); and fat, you know, will not mingle with water. Both these parts of meat, both these food-stuffs, must be acted upon before they can pass from the inside of the alimentary canal, through the epithelium of the mucous membrane, into the blood capillaries.

Besides meat we eat bread. Bread is chiefly composed of **starch**; but besides starch we find in it a substance containing nitrogen, exceedingly like the proteid matter of muscle or of blood.

Potatoes contain a very great deal of starch with a very small quantity of proteid matter; and nearly all the vegetables we eat contain starch, with more or less proteid matter.

Then we generally eat more or less sugar, either as such or in the form of sweet fruits. We also take salt with our meals, and in almost everything we eat, animal or vegetable, meat, bread, potatoes or fruit, we swallow a quantity of mineral substances, that is, various kinds of salts, such as potash, lime, magnesia, iron, with sulphuric, hydrochloric, phosphoric, and other acids.

In everything on which we live we find one or more of the following food-stuffs:—Proteid matter,

starch or sugar, and fat, together with certain minerals and water. It is on these we live: any article which contains either proteid matter, or starch, or fat, is useful for food. Any article which contains none of them is useless for food, unless it be for the sake of the minerals or water it holds.

We are not obliged to eat all these food-stuffs. Proteid matter we must have always. It is the only food-stuff which contains nitrogen. It is the only substance which can renew the nitrogenous proteid matter of the blood and so the nitrogenous proteid matter of the body.

We might indeed manage to live on proteid matter alone, for it contains not only nitrogen but also carbon and hydrogen, and out of it, with the help of a few minerals, we might renew the whole blood and build up any and every part of the body. But, as you will learn hereafter, it would be uneconomical and unwise to do so. Starch, sugar, and fats, contain carbon and hydrogen without nitrogen; and hence, if we are to live on these we must add some proteid matter to them.

50. Of these food-stuffs, putting on one side the minerals, sugar (of which, as you know, there are several kinds, cane sugar, grape sugar, and the like) is the only one which is really soluble, and will pass readily by osmosis through thin membranes (see p. 84). If you take a quantity of white of egg, or blood serum, or meat, or fibrin, or a quantity of starch boiled or unboiled, or a quantity of oil or fat, place it in a bladder, and immerse the bladder in pure water, you will find that none of it passes through the bladder into the water outside, as sugar or salt would do. In the same way a quantity of meat, or of starch, or of fat, placed in your alimentary canal, would never get through the membrane which separates the inside of the canal from the inside of the capillaries, and so would remain perfectly useless as food unless something were done to it. While the food is simply inside the alimentary canal, it is really outside your body. It can only be said to be inside your body when it gets into your blood.

In the things we eat, moreover, these food-stuffs are mixed up with a great many things that are not food-stuffs at all; they are packed away in all manner of little cases, which are for the most part no more good for eating than the boxes or paper in which the sweetmeats you buy are wrapped up. The food-stuffs have to be dissolved out of these boxes and packing.

The juices secreted by the glands of which we have been speaking, dissolve the food-stuffs out of their wrappings, act upon them so as to make them fit to pass into the blood, and leave all the wrappings as useless stuff which passes out of the alimentary canal without entering into the blood, and therefore without really forming part of the body at all.

This preparation and dissolving of food-stuffs is called digestion.

Different food-stuffs are acted upon in different parts of the alimentary canal.

The saliva of the mouth has a wonderful power of **changing starch into sugar**. If you take a mouthful of boiled starch, which is thick, sticky, pasty, and tasteless, and hold it in your mouth for a few moments, it will become thin and watery, and will taste quite sweet, because the starch has been changed into sugar. Now sugar, as you know, will readily pass through membranes, though starch will not.

The gastric juice in the stomach does not act much on starch, but it **rapidly dissolves all proteid matters**.

If you take a piece of boiled meat, put it in some gastric juice and keep the mixture warm, in a very short time the meat will gradually disappear. All the proteid matter will be dissolved, and only the wrappings of the muscular fibre and the fat be left. You will have a solution of meat—a solution, moreover, which, strange to say, will easily pass through membranes, and is therefore ready to get into the blood.

The pancreatic juice and the juice secreted by the intestine act both on starch as saliva does, and on proteids very much as gastric juice does.

51. The bile and the pancreatic juice together act upon all fats in a very curious way.

You know that if you shake up oil and water together, though by violent shaking you may mix them a good deal, directly you leave off they separate again, and all the oil is seen floating on the top of the water. If, however, you shake up oil with pancreatic juice and bile, the oil does not separate. You get a sort of creamy mixture, and will have to wait a very long time before the oil floats to the top. Milk, you know, contains fat, the fat which is generally called butter. If you examine milk under the microscope, you will find that the fat is all separated into the tiniest possible drops. So also, when you shake up oil or butter, or any other fat, with bile and pancreatic juice, you will find on examination that the fat or oil is all separated into the tiniest possible drops. What is the purpose of this?

If you look at the inside of the small intestine of any animal, you will find that it is not smooth and shiny like the outside of the intestine, but shaggy, or, rather, velvety. This is because the mucous membrane is crowded all over with little tags, like very little tongues, hanging down into the inside of

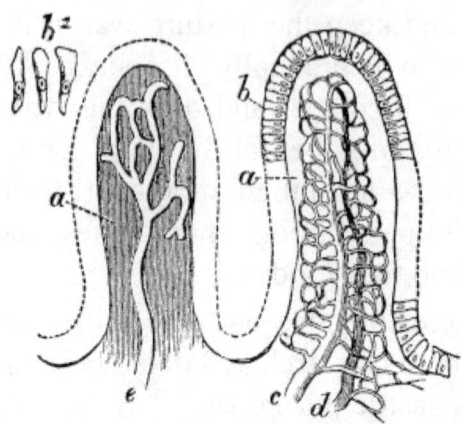

Fig. 18.—Semi-diagrammatic View of Two Villi of the Small Intestines. (Magnified about 50 diameters.)

a, substance of the villus; b, its epithelium, of which some cells are seen detached at b2; c d, the artery and vein, with their connecting capillary network, which envelopes and hides e, the lacteal which occupies the centre of the villus and opens into a network of lacteal vessels at its base.

the intestine. These are called **villi**; they are not unlike the papillæ of the skin (Fig. 15), if you suppose all the epidermis stripped except the bottom row of cells (*d*), and the papilla itself pulled out a good deal. Fig. 18 is a sketch to illustrate the structure of a **villus**. The epithelium (*b*), you see, is made up of a single row of cells. Beneath the epithelium, just as in the papilla of the skin, is a network of blood capillaries, shown, for convenience, in the right-hand villus only. But besides the blood capillaries, there is in each villus, what there is not in a papilla of the skin, another capillary (shown, for convenience, in the left-hand villus only) which does not contain blood, which is not connected with any artery or with any vein, but which begins in the villus. This is a **lacteal**. I have said nothing of these at present. In most parts of the body we find, besides blood capillaries, fine passages very much like capillaries, except that they contain a colourless fluid instead of blood, and do not branch off from any larger vessels like arteries. They seem to start out of the part in which they are found, like the roots of a plant in the soil. But though unlike blood capillaries in not branching off from larger trunks, they resemble capillaries in joining together to form larger trunks corresponding to veins, and the colourless fluid flows from the fine capillary channels towards these larger trunks. This colourless fluid is called **lymph**; it is very much like blood without the red corpuscles, and the channels in which it flows are called **lymphatics**.

The lymphatics from nearly all parts of the body join at last into a great trunk called the **thoracic duct**, which empties itself into the great veins of the neck, as is shown in the diagram, Fig. 6, *Lct.*, *Ly.*, *Th. D.*

Now, many of the lymphatics start from the innumerable villi of the intestine, and are there called **lacteals** (Fig. 6, *Lct.*); so that lacteals may be said to be those lymphatics which have their roots in the villi of the intestine.

But what has all this to do with the digestion of fat? Lacteal means **milky**, and the lymphatics coming from the villi are called lacteals because, when digestion is going on, the fluid in them, instead of being transparent as in the rest of the lymphatics, **is white and milky**. Why is it thus white and milky? Because it is crowded with minute particles of fat, and those minute particles of fat come from the inside of the intestine. They are the same minute particles into which the bile and pancreatic juice have divided the fat taken as food. We know this because when no fat is eaten the lacteals do not get milky; and when for any reason bile and pancreatic juice are prevented from getting into the intestine, though ever so much fat be eaten, it does not get into the lacteals at all, it remains in the intestine in great pieces, and is finally cast out as useless.

52. This, then, is what becomes of the food-stuffs:—

The fats are broken up by the bile and pancreatic juice into minute particles. These minute particles, we do not exactly know how, pass through the epithelium of the villus into the lacteal vessels, from the lacteals into the thoracic duct, and from the thoracic duct into the vena cava. Thus the fats we eat get into the blood.

The starch is changed into sugar in the mouth by saliva, and in the intestine by the pancreatic juice; but sugar passes readily through membranes, and so slips into the blood capillaries of the walls of the alimentary canal. Thus all the sugar we eat, and all the goodness of the starch we eat, pass into the blood.

The proteids are dissolved in the stomach by the gastric juice, and what passes the stomach is dissolved in the intestine, dissolved in such a way that it can pass through membranes; and thus proteids pass into the blood.

Probably some of the sugar and proteids pass into the lacteals as well.

The minerals are dissolved either in the mouth, or in the stomach, or in the intestine, and pass into the blood.

And water passes into the blood everywhere along the whole length of the canal.

When we eat a piece of bread, while we are chewing it in our mouth it is getting moistened and mixed with saliva. Part of its starch is thereby changed into sugar, and all of it is softened and loosened. Passing into the stomach, some of the proteids are dissolved out by the gastric juice, and pass into the blood, and all the rest of the bread breaks up into a pulpy mass. Passing then into the intestine, what is left of the starch is changed by the pancreatic juice into sugar, and is at once drained off either into the lacteals or straight into the blood. In the intestine what remains of the proteids is dissolved, till nothing is left but the shells of the tiny chambers in which the starch and proteids were stored up by the wheat-plant as it grew.

When we eat a piece of meat, it is torn into morsels by the teeth and well moistened by saliva, but suffers else little change in the mouth. In the stomach, however, the proteids rapidly vanish under the action of the gastric juice. The morsels soften, the fibres of the muscle break short off and come asunder; the fat is set free from the chambers in which it was stored up by the living ox or sheep, and, melted by the warmth of the stomach, floats in great drops on the top of the softened pulpy mass of the half-digested food. Rolled about in the stomach for some time by the contraction of the muscles which help to form the stomach walls, losing much of its proteids all the while to the hungry blood, the much-changed meat is squeezed into the intestine. Here the bile and the pancreatic juice, breaking up the fat into tiny particles, mix fat, and broken meat, and empty wrappings, and salts, and water, all together into a thick, dirty, yellowish cream. Squeezed along the intestine by the contraction of the muscular walls, the goodness of this cream is little by little sucked up. The fat goes drop by drop, particle by particle, into the lacteals, and so away into the blood. The proteids, more and more dissolved the further they travel along the canal, soak away into blood-vessel or into lacteal. The salts and the water go the same way, until at last the digested meat, with all its goodness gone, with nothing left but indigestible wrappings, or perhaps as well some broken bits of fibre or of fat, is cast aside as no longer of any use.

Thus all food-stuffs, not much altered, with all their goodness unchanged, pass either at once into the

blood, or first into the lacteals and then into the blood, and the useless wrappings of the food-stuffs are cast away.

While we are digesting, the blood is for ever rushing along the branches of the aorta, through the small arteries and capillaries of the stomach and intestine, along the branches of the portal vein, and so through the liver back to the heart; and during the few seconds it tarries in the intestine, it loads itself with food-stuffs from the alimentary canal, becoming richer and richer at every round. While we are digesting, the thoracic duct is pouring, drop by drop, into the great veins of the neck the rich milky fluid brought to it by the lacteals from the intestine, and as the blood sweeps by the opening of the thoracic duct on its way down from the neck to the heart, it carries that rich milky fluid with it, and the heart scatters it again all over the body.

Thus the blood feeds on the food we eat, and the body feeds on the blood.

53. But if the blood is thus continually being made rich by things, it must also as continually be getting rid of things. The things with which it parts are not, however, the same as those which it takes. The blood, as we have said, is fuel for the muscles, for the brain, and for other parts of the body. These burn the blood, burn it with heat but without light. But, as you have learnt from your Chemistry Primer, Art. 4, burning is only change, not destruction; in burning nothing is lost. If the muscle burns blood, it burns it into something; that something, being already burnt, cannot be burnt again, and must be got rid of.

Into what things does the body burn itself while it is alive?

I have already said that if you were to take a piece of meat or some blood, and dry it and burn it, you would find that it was turned into four things—water, carbonic acid, ammonia, and ashes. The body is made up of nitrogen, carbon, hydrogen, and oxygen, with sulphur, phosphorus, and some other elements. The nitrogen and hydrogen go to form ammonia; the hydrogen, with the oxygen of combustion, forms water; the carbon, carbonic acid; the phosphorus, sulphur, and other elements go to form phosphates, sulphates, and other salts.

In whatever way the body be oxidized, whether it be rapidly burnt in a furnace, whether it be slowly oxidized after death, as when it moulders away either above ground or in the soil, whether it be quickly oxidized by living arterial blood while still alive—in all these several ways the things into which it is burnt, into which it is oxidized, are the same. Whatever be

the steps, the end is always water, carbonic acid, ammonia, and salts.

These are the things which are always being formed in the blood through the oxidation of the body, these are the things of which the body has always to be getting rid.

In addition to the water which comes from the oxidation of the solids of the body, we are always taking in an immense quantity of water; partly because it is absolutely necessary that our bodies within should be kept continually moist, partly because food cannot pass into the blood except when dissolved in water, and partly because we need washing inside quite as much as outside; if we had not, so to speak, a stream of water continually passing through our bodies to wash away all impurities, we should soon be choked, just as an engine is choked with soot and ashes if it be not properly cleaned. We have, then, to get rid daily of a large quantity of washing water over and above that which comes from the burning of the hydrogen of our food.

We have already seen that a great deal of the carbonic acid goes out by the lungs at the same time that the oxygen comes in. A large quantity of water escapes by the same channel. You very well know that however dry the air you breathe, it comes out of your body quite wet with water.

We have also already seen how the blood secretes sweat into the sweat-glands, and so on to the skin. Perspiration is little more than water with a little salt in it. The skin, therefore, helps to purify the blood through the sweat-glands, by getting rid of water with a little salt. You must remember that a great deal of water passes away from your skin without your knowing it. Instead of settling on the skin in drops of sweat, it passes off at once as vapour or steam. Some carbonic acid also makes its way from the blood through the skin.

54. It only remains for us to inquire, In what way does the blood get rid of the ammonia and the rest of the saline matters that do not pass through the skin?

These are secreted from the blood by the kidney, dissolved in a large quantity of water in the form of **urine**.

What is the kidney? You will learn more about this organ by and by. Meanwhile it will for our present purpose be sufficient to say that a kidney is a bundle of long tubular glands, not so very unlike sweat-glands, all bound together into the rounded mass whose appearance is familiar to you. **Into these glands the blood secretes urine just as it secretes sweat into the sweat-glands.** The glands themselves unite into a common tube or duct which carries the urine into the receptacle called the urinary bladder, from whence it is cast out when required.

What is urine? Urine is in reality water holding in solution several salts, and in particular containing a quantity of ammonia. The ammonia in urine is generally in a particular condition, being combined with a little carbonic acid, in the form of what is called **urea**. If urea is not actually ammonia, it is at least next door to it.

The three great channels, then, by which the blood purifies itself, by which it gets rid of its waste, are the lungs, the kidneys, and the skin. Through the lungs, carbonic acid and water escape; through the kidneys, water, ammonia in the shape of urea, and various salts; through the skin, water and a few salts. As the blood passes through lung, kidney, and skin, it throws off little by little the impurities which clog it, one at one place, another at another, and returns from each purer and fresher. The need to get rid of carbonic acid and to gain a fresh supply of oxygen is more pressing than the need to get rid of either ammonia or salts. Hence, while all the blood which leaves the left ventricle has to pass through the lungs before it returns to the left ventricle again, only a small part of it passes through the kidneys, just enough to fill at each stroke the small arteries leading to those organs. The blood craves for great draughts of oxygen, and breathes out great mouthfuls of carbonic acid, but is quite content to part with its ammonia and salts in little driblets, bit by bit.

The three channels manage between them to keep the blood pure and fresh, working hard and clearing off much when much food or water is taken or much work is done, and taking their ease and working slow when little food is eaten or when the body is at rest.

55. And now you ought to be able to understand how it is that we live on the food we eat.

Food, inasmuch as it can be burnt, is a source of power. In burning it gives forth heat, and heat is power. If we so pleased, we might burn in a furnace the things which we eat as food, and with them drive a locomotive or work a mill; if we so pleased, we might convert them into gunpowder, and with them fire cannon or blast rocks. Instead of doing so, we burn them in our own bodies, and use their power in ourselves.

Food passing into the alimentary canal is there digested; the nourishing food-stuffs are with very little change dissolved out from the innutritious refuse; they pass into and become part and parcel of the blood.

The blood, driven by the unresting stroke of the heart's pump, courses throughout the whole body, and in the narrow capillaries bathes every smallest bit of almost every part. Kept continually rich in combustible material by frequent supplies of food, the blood as well at every round sucks up oxygen from the air of the lungs; and thus arterial blood is ever carrying to all parts of the body, to muscle, brain, bone, nerve, skin, and gland, stuff to burn and oxygen to burn it with.

Everywhere oxidation, burning, is going on, in some spots or at some times fiercely, in other spots or at other times faintly, changing the arterial blood rich in oxygen to venous blood poor in oxygen. From most places where oxidation is going on, the venous blood goes away hotter than the arterial which came; and all the hot blood mingling together and rushing over the whole body keeps the whole body warm. Sweeping as it continually does through innumerable little furnaces, the blood must needs be warm. This is why **we** are warm. But from some places, as from the skin, the venous blood goes away cooler than the arterial which came, because while journeying through the capillaries of the skin it has given up much of its heat to whatever is touching the skin, and has also lost much heat in turning liquid perspiration into vapour. This is why so long as we are in health we never get hotter than a certain degree of temperature, the so-called blood-heat, 98° Fahr., and why we make warm the clothes which we wear and the bed in which we sleep.

Everywhere oxidation is going on, oxidation either of the blood itself or of the structures which it bathes, and whose losses it has to make good. Everywhere change is going on. Little by little, bit by bit, every part of the body, here quickly, there slowly, is continually mouldering away and as continually being made anew by the blood. Made anew according to its own nature. Though it is the same blood which is rushing through all the capillaries, it makes different things in different parts. In the muscle it makes muscle; in the nerve, nerve; in the bone, bone; in the glands, juice. Though it is the same blood, it gives different qualities to different parts: out of it one gland makes saliva, another gastric juice: out of it the bone gets strength, the brain power to feel, the muscle power to contract.

When the biceps muscle contracts and raises the arm, it does work. The power to do that work, the muscle got from the blood, and the blood from the food. All the work of which we are capable comes, then, from our food, from the oxidation of our food, just as the power of the steam-engine comes from the oxidation of its fuel. But you know that in the steam-engine only a very small part of the power, or energy, as it is called, of the fuel goes to move the wheel. By far the greater part is lost in heat. So it is with our bodies: all the force we can exert with our bodies is but a small part of the power of our food; all the rest goes to keep us warm.

Visiting all parts of the body, rebuilding and refreshing every spot it touches, the blood current also carries away from each organ the waste matters of which that organ has no longer any use. Just as each part or organ has different properties and different work, so also is the waste of each not exactly the same, though all are alike inasmuch as they are all the results of oxidation. The waste of the muscle is not exactly the same as the waste of the brain or of the liver. Possibly the waste things which the blood bears from one organ may be useful to another, and so be made to do double work, just as the tar which the gasworks throw away makes the fortune of the colour manufacturer.

Be this as it may, the waste products of all parts, travelling hither and thither in the body, come at last to be brought down to very simple things, with all their virtue gone out of them, with all, or all but all, their power of burning lost, fit for nothing but to be cast away, come at last to be urea or ammonia, carbonic acid, and salts. In this shape, the food, after a longer or shorter sojourn in the body, having done its work, having built up this or that part, having helped the muscle to contract or the liver to secrete, having by its burning given rise to work or to heat, goes back powerless to the earth and air from which it came. And so the tale is told.

56. One other matter we have to note before we have given the full answer to the question why we move.

We have seen that we move by reason of our muscles contracting, and that in a general way a muscle contracts because a something started in the brain by our will passes down from the brain through more or less of the spinal cord, along certain nerves till it reaches the muscle. It is this something, which we may call a **nervous impulse**, which causes the muscle to contract.

But what leads us to exercise our wills? What starts the nervous impulse?

All the nerves in the body do not end in muscles. Many of them end, for instance, in the skin, in those papillæ of which I spoke a little while ago. These nerves cannot be used for carrying nervous impulses from the brain to the skin. By an effort of the will you can make your muscles contract; but try as much as you can, you cannot produce any change in your skin.

What purpose do these nerves serve, then? If you prick or touch your finger, you feel the prick or touch; you say you have **sensation** in your finger. Suppose you were to cut across the nerves which lead from the skin of your finger along your arm up to your brain. What would happen? If you pricked or touched your finger, you would not feel either prick or touch. You would say you had lost all sensation in your finger. These nerves ending in the finger then, have a different use from those ending in the muscle. **The latter carry impulses from the brain to the muscle, and so, being instruments for causing movements, are called motor nerves. The former, carrying impulses from the skin to the brain, and being instruments for bringing about sensations, are**

called sensory nerves. All parts of the skin are provided with these sensory nerves, but not to the same extent. The parts where they abound, as the fingers, are said to be very sensitive; the parts where they are scanty, as the back of the trunk, are said to be less sensitive. Other parts besides the skin have also sensory nerves.

Motor nerves are of one kind only; they all have one kind of work to do—to make a muscle contract. But there are several kinds of sensory nerves, each kind having a special work to do. The several works which these different kinds of sensory nerves have to do are called **the senses**.

The work of the nerves of the skin, all over the body, is called the **sense of touch**. By touch you can learn whether a body is rough or smooth, wet or dry, hot or cold, and so on.

You cannot, however, by touch distinguish between salt and sugar. Yet directly you place either salt or sugar on your tongue you can recognize it, because you then employ sensory nerves of another kind, the nerves which give us the **sense of taste**. So also we have nerves of **smell**, nerves of **hearing**, and nerves of **sight**.

The nerves of touch, where they end, or rather where they begin in the skin, sometimes have and sometimes have not, little peculiar structures attached to them, little **organs of touch**. So also the nerves of taste, and smell, end or rather begin in a peculiar way. When we come to the nerves of hearing and of seeing, we find these beginning in most elaborate and complicated organs, the ear and the eye.

Of all these **organs of the senses** you will learn more hereafter; meanwhile, I want you to understand that by means of these various sensory nerves, we are, so long as we are alive and awake, receiving impressions from the external world, sensations of touch, sensations of roughness and smoothness, of heat and cold, sensations of good and bad odours, sensations of tastes of various kinds, sensations of all manner of sounds, sensations of the colours and forms of things.

By our skin, by our nose, by our tongue and palate, by our ears, and above all by our eyes, impressions caused by the external world are for ever travelling up sensory nerves to the brain; thither come also impressions from within ourselves, telling us where our limbs are and what our muscles are doing. Within the brain these impressions become sensations. They stir the brain to action; and the brain, working on them and by them, through ways we know not of, governs the body as a conscious intelligent will.

FOOTNOTES:

[1] It is unusual for muscles to have two tendons at the same end. Hence the name **biceps**, or "two-headed."

[2] From *pulmo*, **lung**; the artery of the lung.

[3] From *hepar*, **liver**; the vein of the liver.

www.ingramcontent.com/pod-product-compliance
Lightning Source LLC
Chambersburg PA
CBHW071446180526
45170CB00001B/486